Beekeeping in California

The Authors

Eric C. Mussen, Extension Apiculturist, University of California, Davis; editor and reviser of all sections.

Len Foote, Chief, Control and Eradication, Division of Plant Industry, California Department of Food and Agriculture, Sacramento, California.

State Laws Relating to Beekeeping
Bee Diseases
Other Disorders
Pests of Bees

Norman E. Gary, Professor of Entomology and Apiculturist, in the Experiment Station, University of California, Davis.

An Observation Beehive

Harry H. Laidlaw, Emeritus Professor of Entomology and Apiculturist, Retired, University of California, Davis.

Maintaining Genetic Stock

Robbin W. Thorp, Professor of Entomology and Apiculturist, in the Experiment Station, University of California, Davis.

Pollinating Crops with Honey Bees
Sources of Nectar and Pollen

Lee H. Watkins, Emeritus Apiarist (deceased), University of California, Davis.

Beekeeping in California

Front and back cover photos by Paul Rosenfeld

Contents

For information about ordering this publication, write to:

Publications
Division of Agriculture and Natural Resources
University of California
6701 San Pablo Avenue
Oakland, California 94608-1239

or telephone (415) 642-2431

Publication 21422
ISBN 0-931876-79-6
Library of Congress Catalog Card Number 87-71574

Printed in the United States of America.

To simplify information, trade names of products have been used. No endorsement of named products is intended, nor is criticism implied of similar products which are not mentioned.

Issued in furtherance of Cooperative Extension work, Acts of May 8 and June 30, 1914, in cooperation with the U.S. Department of Agriculture. Kenneth R. Farrell, Director of Cooperative Extension, University of California.

5m-pr-12/87-HS/FB Formerly Publication 4026

Foreword

California beekeeping operations range in size from commercial enterprises of 1,000 to 15,000 colonies owned by a single individual to just a couple of colonies in a beekeeper's backyard. Each of these colonies has the same basic needs that must be satisfied to ensure adequate strength and productivity. This publication describes the fundamentals of keeping bees in California and discusses the differences between commercial and noncommercial approaches. However, many procedures mentioned are not described fully, and inexperienced beekeepers are advised to read other references for details.

Beekeeping in California

The first known evidence that early man robbed honey from bees is a primitive drawing on a cave wall in eastern Spain dating from 7000 B.C. Throughout recorded history honey's importance as a food and as medicine has been realized. English settlers brought the honey bee (*Apis mellifera* L.) to North America in about 1622. Thomas Jefferson, in his *Notes on the State of Virginia*, observed that American Indians called the honey bee "the white man's fly." In California, honey bees were introduced in 1853 by Christopher A. Shelton, who established an apiary of 12 colonies just north of San Jose. Of the 12, only one survived, but it cast three swarms that summer and by 1858 there were at least 150 colonies directly descended from the Shelton hive.

California's first professional beekeeper, John S. Harbison, imported 57 colonies from Pennsylvania to Sacramento in December 1857; of these 50 survived. He increased them by artificial division to 136 hives and sold 130 for $100 each. Harbison then imported 114 colonies, lost 11, and in 1859 sold nearly $30,000 worth of bees, keeping 138 colonies for himself for the next season. Harbison's sensational success started a "bee rush" to California, and in 1859 and 1860 more than 8,000 colonies were imported from the East Coast via the Isthmus of Panama, the largest long-distance shipment of honey bees ever attempted. In 1869 Harbison moved his bees from Sacramento to the newly discovered sage and wild buckwheat ranges of San Diego County, and by 1873 San Diego County had produced more honey than any other county in California. By 1876, Harbison had 3,750 colonies of bees in 12 apiaries and was recognized as the largest honey producer in the world. Since that time, California has been one of the nation's principal honey-producing states.

Twentieth century beekeeping has its own unique problems, mostly the result of increased urbanization and the consequence that nectar sources are widely scattered. Fortunately, small numbers of colonies often can do well in or near cities because of the diversity of flowering plants within their flight range. The amateur beekeeper may often profit from this fact.

Value of the industry

From an economic standpoint, honey bees make their greatest contribution to California agriculture as pollinators of commercial crops. The following crops, realizing more then $1 billion annually, require bee pollination: alfalfa seed, almonds, most apples, avocados, Bartlett pears, bushberries, cherries, cucumbers, flaxseed, kiwi, Ladino clover seed, melons, plums, prunes, pumpkins, rape seed, safflower seed, squash, sunflower seed, tangelos, tangerines, 22 vegetable seeds, and flower seeds. Many of these crops, as well as ornamental plants, are grown in home or community gardens where bee pollination is equally as essential for producing seeds, fruits, or vegetables. Bees also pollinate weeds, which provide food for wild birds and mammals and prevent erosion of watersheds and wilderness areas. Bee pollination has an enormous impact on our diets and on the stability of our environment.

Some California beekeepers specialize in producing queens and packaged bees for sale for starting new colonies or requeening functioning units. More than 450,000 packages (each with a queen) are shipped annually to northern honey producers to restock hives that are emptied in winter. Another 150,000 queens are sold for installation in overwintered hives or to quick start recently divided colonies. The demand for queens and packages produced by reputable bee breeders is great, because commercially produced stocks are most likely to be mild-tempered, good producers.

Commercial honey production in California varies from good to poor, depending upon weather conditions. Most beekeepers equipped to move bees take at least a portion of their bees to potential honey-producing areas each year. When the nectar flow is heavy in certain areas, thousands of hives may be moved in. The California honey crop averages about 20 million pounds per year, not enough to meet the state's consumer demand. Honey packing and importing are important segments of the California beekeeping industry.

Hobby beekeepers often keep their bees in one location. Honey crops can be quite good in coastal, urban, or suburban areas where weeds, trees, flowers, and shrubs are apt to bloom most of the year. However, where lack of rainfall creates long nectar and pollen dearths during summer, colonies must be examined often and frequently must be fed to avoid malnutrition or starvation.

Keeping bees for fun and profit

Interest in keeping bees increased during the 1970s, largely because of the conviction that natural foods are preferable to processed foods. Thus, honey appeared to be an ideal substitute for sugar. The aesthetic values of beekeeping are also often important. Observing bee behavior at the hive (regulation of population size and use of space within the hive) or outside the hive (foraging for water, nectar, pollen, and propolis, plus pollination ecology) is a leisurely way to relax and enjoy life. The rewards for successful colony management and the consequences of unsuccessful management are apparent.

Every beekeeper can realize a profit, if that is a goal. Locally produced honey usually sells quickly at, or slightly above, supermarket prices when hobby beekeepers advertise its availability. Beekeepers who wish to augment their incomes substantially must operate between 50 and 500 colonies. Most commercial beekeepers operate between 1,000 and 2,000 colonies, often with one or more permanent helpers and part-time employees for extracting honey or shaking bees. Most commercial beekeepers in California engage in crop pollination during the year, from which they earn a significant portion of their annual income.

Becoming a Beekeeper

Persons considering keeping bees can learn through self education and experience. Classes and short courses in beekeeping are also helpful, and many good books and other literature are available (see *References*). However, no amount of reading can substitute for actual experience with colonies. Local beekeeping clubs often willingly share information, and many will show beginners how to manage a colony and what to expect through the year. Those seeking financial profit should apprentice themselves to a commercial beekeeper for a year or two to learn the ropes. Names of beekeepers often are available from county agricultural commissioners, county farm advisors, local police and fire departments, and animal control units.

If you decide to start keeping bees:

(1) Check state and local laws for possible restrictions on keeping bees (see below).

(2) Determine your sensitivity to bee stings—your doctor can test this.

(3) Purchase, assemble, and paint standard-size equipment well in advance of the anticipated arrival of the bees.

(4) Locate your apiary close to home, away from pedestrians and auto traffic, and where the bees will not disturb people or livestock.

(5) Provide a permanent, functional watering device if a natural source of water is not readily available.

(6) Avoid placing hives in areas where pests (ants, skunks, bears) or poisonous plants (California buckeye, locoweed, corn lily, or death camas) may damage the colony.

(7) Protect your bees from strong winds and hot summer sunshine.

Beekeeping organizations

Groups of California beekeepers have been meeting for nearly a century. At the state level, the California State Beekeepers' Association represents the interests of the commercial beekeepers, although a number of noncommercial beekeepers attend their annual meeting in November.

Many local clubs have formed on a county basis. These clubs tend to represent commercial, hobby, or mixed interests depending upon the makeup of the group. Club members know best how to keep bees in their local areas, and they are willing to share that information.

Names and addresses of contact persons for these organizations tend to change over time. However, the Cooperative Extension apiculturist, county agricultural commissioners, or Cooperative Extension farm advisors should be able to steer you to a local group.

STATE LAWS RELATING TO BEEKEEPING

California laws regulating beekeeping are enforced by county agricultural commissioners and provide the basis for an effective apiary inspection program that helps beekeepers protect honey bee colonies from disease, pesticide damage, and theft.

Excerpts from the California Agricultural Code relating to bees and apiary inspection can be purchased from: Office Services, California Department of Agriculture, 1220 N Street, Sacramento, CA 95814; (916) 445-8164. Beekeeping in some localities is also governed by city or county ordinances. Beekeepers should consult local authorities about this.

Apiary registration. All apiaries must be registered each January with the agricultural commissioner of the county in which the colonies are located. Registration fee is $10 and involves listing the location of each apiary and the number of colonies at each location. Newly acquired apiaries and apiaries brought from out of state must be registered within 30 days of establishment.

Apiary movements and identification. Details of laws pertaining to movement and identification of apiaries can be obtained from county agricultural commissioners or Supervisor of Apiary Projects, California Department of Agriculture, 1220 N Street, Sacramento, CA 95814.

Apiary assessment. Resident and nonresident beekeepers operating 40 or more colonies in California are required to pay an annual assessment fee on their colonies. The rate has varied for several years, so the Supervisor of Apiary Projects (address above) should be contacted for current rates.

The Colony

A colony of bees consists of a queen, worker bees, drones, and various stages of brood (immature bees) living together as a social unit. There are between 10,000 and 50,000 bees in a colony. The brood nest is spherical in shape, increasingly filling more cells in each comb and covering more combs as it expands in size. Partially digested pollen, called bee bread, is stored adjacent to cells containing brood. Honey or nectar is stored around the outer edges of, and above, the brood nest.

A honey bee egg looks like a tiny grain of white rice standing on end, centered at the base of a cell. To facilitate seeing eggs and other larval stages, shake or gently brush the bees off the comb (use a bee brush) and stand with your back to the sun. Tilt the comb so that the light shines directly into the cells. With a little experience it is not difficult to recognize larval bees or to distinguish capped brood (pupae) from capped honey (ripened honey covered by a thin layer of wax). See Plate I.

The queen bee

Each colony normally has only one queen (Plate I), which is the only bee in the colony capable of fertilizing the eggs she lays. The queen bee develops from a fertilized egg that hatches 3 days after being laid. Nurse bees, a class of worker bee, feed developing queen larvae a special diet consisting mostly of the royal jelly that they secrete from their glands. This special diet shortens the time spent to reach maturity to 16 days, compared with 21 days for the worker bee and 24 for the drone. The result is a bee larger than any others, with fully developed ovaries and a very large abdomen. The queen lacks the specialized body parts of worker bees that help them accomplish their tasks. The queen's task is to produce bees and the constant diet of royal jelly fed to an adult queen supplies the nutrients necessary for development of the large ovaries that swell the abdomen.

The queen is reared in a large cell resembling a peanut shell that hangs vertically from the comb (Plate I), and about 10 days after emerging she becomes sexually mature. The virgin queen takes one or more brief mating flights during which she mates with 10 to 20 drones to ensure complete filling of the spermatheca. Large amounts of sperm are necessary, since the queen will be laying more than 1,000 eggs a day for many months and will never mate again. The queen begins laying eggs shortly after mating.

Even though the queen has a larger thorax, longer abdomen, and less hair than the workers, she can be very difficult to find in a populous colony. Clipping and marking the queen is worth much more than the few cents it costs when she has to be located in the colony. To ensure the potential for having a populous and productive colony, beekeepers should requeen their colonies annually with young vigorous queens (see *Maintaining Genetic Stock*).

The drone bee

At their peak population (early summer), drones rarely exceed 600 per colony. Their sole function, as male bees, is to mate with the queen. When virgin queens are no longer being produced (in the fall), the drones are forced out of the colony to die of starvation, and no drones are reared until the following spring.

Drones develop from unfertilized eggs that hatch 3 days after they are laid. Nurse bees feed the developing larvae royal jelly, honey, and pollen over a 7-day period; the cells are then covered with air-permeable wax (capped). A drone pupa is longer than a worker pupa; thus, its capping is raised above the surface of the comb. This is especially apparent if the drone is reared in a worker cell, where the capping rises way above the capped worker brood and sometimes is referred to as a "bullet." The drone emerges 24 days after the egg is laid and spends the next 10 days maturing sexually and learning to fly. A drone must be fed by worker bees from the time he emerges until the day he dies of old age (about 5 weeks after emerging) or immediately after mating with a virgin queen.

The drone can be distinguished from the workers by its large size, blocky shape, and very large eyes which cover most of his head. He makes more noise when flying than does the worker, but he is harmless because he has no sting.

The worker bee

All the rest of the bees in the colony are workers. The worker bee develops from a fertilized egg that hatches 3 days after it is laid. Nurse bees feed the developing larva royal jelly, honey, and pollen during the next 5 to 6 days, then cap the cell. Each larva spins a cocoon and changes to a prepupa, then a pupa. The pupa is not physically active, but undergoes extensive chemical and structural changes that convert it into a functioning adult. (Adult workers are always female.) On the 21st day after the egg has been laid, the adult chews through her wax cap and emerges from the cell to groom herself and to start eating honey and pollen. Her exoskeleton hardens and she is ready to begin her many chores.

The workers, endowed with specialized body parts to accomplish their tasks, supply all the labor of the colony. Young worker bees clean cells, feed larvae (through food glands in the workers' heads), remove debris from the hive, evaporate water from nectar to produce honey, secrete wax (through wax glands in their abdomens), build the comb, guard the colony (by means of their inbuilt chemical alarm system), and ventilate the hive. When they are about 3 weeks old, worker bees begin to forage for water and nectar, carrying their finds in a honey sac. Worker bees live only 6 weeks or so during periods of active brood rearing and foraging, but they can survive for several months over winter.

Annual colony cycle

The yearly cycle of the colony begins in January when the queen starts to lay eggs in response to an increasingly longer day. The population of the brood nest continues to increase as long as there is an adequate supply of honey and pollen stored in the hive. Fresh pollen collected by foraging workers from early spring flowers signals the beginning of a great increase in brood rearing. Newly emerged workers, well suited for producing royal jelly and wax, accelerate the population explosion.

Rapidly becoming filled with bees, brood, and food, the hive may become congested. Congestion often leads to swarming, especially when an old queen is in residence (see *Managing Bees*). Worker bees in colonies manipulated to discourage swarming collect nectar and pollen in surplus of their immediate needs. This surplus is stored for use when food is not available in the field (dearth). Honey bees store more honey than they need for a year, it nectar is abundant. This excess is the beekeeper's reward for proper colony management.

Nectar and pollen become scarce at the end of summer. Brood rearing decreases markedly. Drones are evicted and the worker population begins to decline. Foraging bees collect extra propolis to close up hive entrances for the winter. The bees become much less active as cool weather sets in. In areas where temperatures fall below 57°F, the bees cluster or form a large ball. Bees in the center of the cluster eat honey and produce heat; bees on the outside of it act as insulation, keeping the heat in the cluster. The rest of the hive and combs not in contact with the cluster are nearly as cold as the outside air. The cluster, moving over the combs and consuming stored honey and pollen, slowly approaches the cover of the hive. In January, the bees increase the temperature in the center of the cluster to around 95°F, the temperature required to rear brood. The necessity of having an abundant supply of stored honey and pollen is readily apparent. Bees wintering in central and southern California frequently are not confined by cold weather and tend to fly much of the year, thereby requiring a great deal of stored honey if nectar is not available to foragers.

Choosing Bees

Most bees reared in California today are Italian or "yellow" bees. Italian bees are noted for good wintering and for extensive brood rearing, which can be beneficial before a good honey flow or detrimental during a honey dearth. The Caucasian or "dark" race of bees is preferred by some beekeepers because these bees tend to be calmer and more gentle when examined. However, Caucasians tend to collect and liberally distribute large amounts of propolis (or bee glue) throughout the hive, thereby making it more difficult to pry the hive components apart.

It is sound practice to purchase commercial queens and bees from reputable breeders. Avoid spending extra money on exotic strains and expensive hybrids until enough experience has been acquired to ensure persistence of your colonies from year to year. Bees can be obtained by: (1) buying a colony in a hive, (2) buying a nucleus, (3) purchasing a queen and bulk bees in a package, and (4) hiving a swarm.

Buying a colony in a hive

This is the easiest way to get started. With the bees already in the hive, it is not necessary to hive them (put them into a hive), as must be done with packaged bees, nuclei, or swarms. *Before it is purchased, the colony should be inspected by the county bee inspector to be sure it is free of disease.*

Buying a nucleus

A nucleus generally consists of three to five frames with bees (6,000 to 10,000 bees), including a laying queen and her brood. Nuclei are transported to hives in mini-hive boxes. Frames and adhering bees are transferred from each nucleus, in identical sequence, into the center of a hive body and surrounded by empty drawn combs or frames with foundation. If stored food is in short supply, the bees should be fed (see *Feeding Bees*).

Buying and installing packages

Packaged bees consist of wire-screen cages in which are confined 2 or 3 pounds (7,000 to 10,000) of worker bees, a queen in a separate small container, and a feeder can of a syrup that is made up of equal parts of water and sugar. Packages can be purchased from a beekeeper or from a bee supply company. Order packages to arrive in spring, 2 to 3 months before the principal nectar-producing plants bloom in your neighborhood. This gives the bees time to build a population large enough to take full advantage of the nectar and pollen when it becomes available.

When the packaged bees arrive, sugar syrup should be sprayed or shaken onto the bees through the screen sides of the package. Give them all they will consume, but not so much that they become stuck together in a mass. The bees can be installed immediately into the waiting hive if it is a cool and cloudy day. If it is warm and sunny, store the bees in a cool, dark room until just before dark. Then, after spraying the bees with syrup again, install the bees in the hive.

Packaged bees are best installed by the "direct release" method, as follows:

(1) Loosely obstruct the hive entrance with a small amount of grass.

(2) Remove four central frames from the hive body.

(3) Pry off the board covering the feeder can on the package.

(4) Rap the package on the ground with sufficient force to knock the cluster to the bottom of the package.

(5) Remove the feeder can and the queen cage.

(6) Lightly spray the bees with syrup if you wish.

(7) Invert the package and roll and jounce the bees out through the round hole into the space between the frames in the hive.

(8) Gently spread out the pile of bees on the bottom board with a hive tool as soon as most of the bees have been dumped out of the package.

(9) Spray or dip the queen in sugar syrup, wetting her wings so she cannot fly.

(10) Without gloves, reach down into the hive and pull the screen off the queen cage.

(11) Place the queen cage against a comb or sheet of foundation and watch the queen as she leaves her cage. She should climb down and go around to the back side of the comb.

(12) Carefully replace the four frames to avoid accidentally injuring the queen. The few bees left in the package will rejoin the group in the hive as soon as the bees remove the grass from the entrance.

This procedure brings the beekeeper very close to the bees. However, packaged bees are quite young and momentarily disoriented so they very rarely sting beekeepers while they are being installed from packages.

Hiving a swarm

Hiving a swarm (catching bees and putting them in a hive) is difficult for a beginner without help from an experienced beekeeper. It is the least expensive way to start beekeeping because swarms usually can be obtained free simply by leaving your name with the county agricultural commissioner, fire department, police department, sheriff, farm advisor, or animal control center.

Newly hived bees should not be disturbed for several days, except to refill the syrup feeder. The queen should begin to lay eggs in a week or less and the colony will start its work. Abundant pollen is necessary for the colony to use as food for rearing brood, feeding the queen, and feeding the drones. Pollen normally is collected from nearby flowers, but when pollen and nectar are not available, it becomes necessary to supply colonies with nutrients. (See *Feeding Bees*.)

Catching swarms has inherent drawbacks: (1) The queen has a tendency to swarm and is likely to do it again, (2) the likelihood of collecting inferior stock (even Africanized bees, eventually) is much greater with swarms than with purchased bees, and (3) the possibility exists that the swarm is carrying contaminated honey with it and a bee disease may break out in the colony. Even though swarming and swarms are intriguing, they are not a very good source of bees for your colonies.

Choosing Equipment

Equipment needed for beekeeping is available from national beekeeping supply dealers and their local representatives; consult the yellow pages of the telephone book. You will need:

The hive (figs. 1 and 2)
☐ Two deep hive bodies or equivalent
☐ Bottom board
☐ Cover
☐ 10 frames, with foundation, per hive body
☐ L-shaped metal rabbets or frame rests (optional)
☐ Honey supers, with frames and foundation (optional)
☐ Queen excluder (optional)
☐ Honey extractor (optional)

Bee stock
☐ One 2-pound package with one queen per hive

Personal equipment (fig. 2)
☐ Smoker
☐ Hive tool
☐ Bee veil
☐ Gloves (optional)
☐ White coveralls (optional)

Hive components

Use only standard size 8- or 10-frame hive equipment (dimensions in fig. 3) so that all parts will fit together properly and match commercially manufactured products. The basic unit of the hive is called a

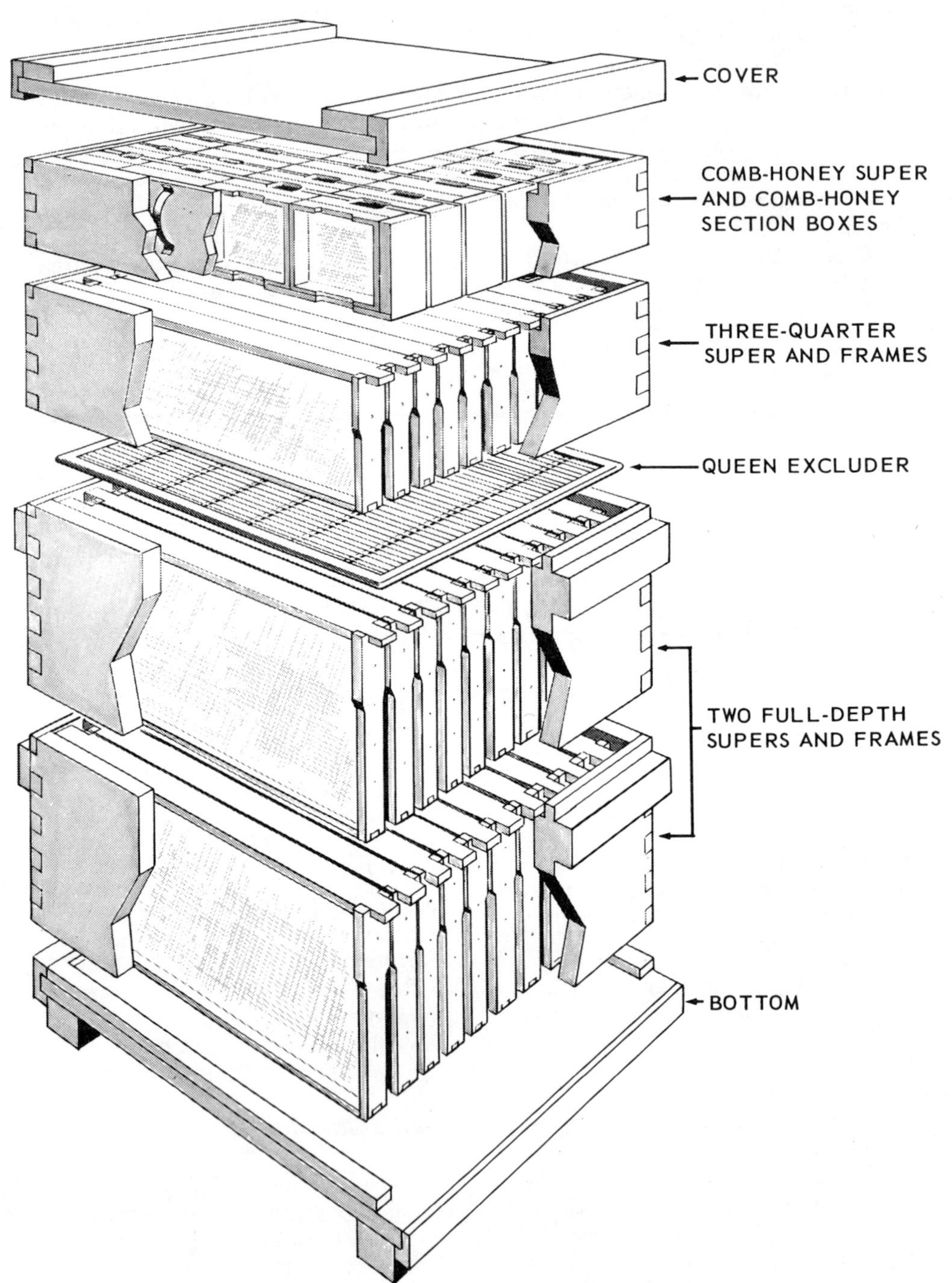

Fig. 1 Beehive detail, showing optional parts used in different types of bee management.

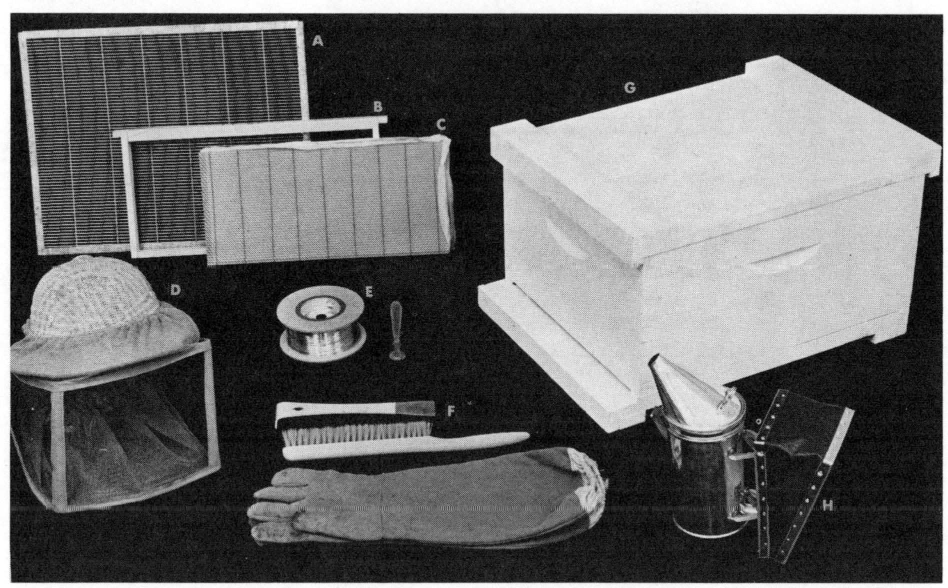

Fig. 2 Basic beekeeping equipment: ***(A)*** **welded wire queen excluder;** ***(B)*** **frame;** ***(C)*** **vertically wired beeswax foundation;** ***(D)*** **hat and wire veil;** ***(E)*** **wire for frames and spur embedder;** ***(F)*** **hive tool, bee brush, and gloves;** ***(G)*** **1-story, 10-frame hive;** ***(H)*** **smoker. All these items are available commercially.**

hive body. A hive body is designed to hold ten (or eight) full-depth (9⅝ inches deep) frames on which the bees will rear their young and store food. A colony eventually will get large enough to cover all ten frames and will require at least two hive bodies to hold all the brood and bees. Boxes placed above the brood nest for storage of honey are called supers. Supers may be full depth, medium, or three-quarter depth (6⅝ inches deep), shallow (5¾ inches deep) or comb honey (4⅝ inches deep). The type of super selected depends upon the type of honey being produced (see *Producing and Marketing Honey*) and the ability of the beekeeper to lift filled boxes.

Each hive body and super originally should be filled with a full set of ten (or eight) frames designed to fit the box. Each frame should have a sheet of the proper size and type foundation firmly attached to the top bar. The bees will follow the pattern embossed on the foundation when building beeswax combs. After all the combs are drawn out fully, one frame can be removed from a ten-frame hive body and the remaining nine frames spaced evenly across the box so that the frames can be manipulated more easily.

Foundation. Commercial foundation is sold in many sizes and types. To assure a proper fit, purchase the frames and foundation from the same supplier. Generally, there are two types of foundation: wired and plastic. Wired is designed for strength and is used for brood combs

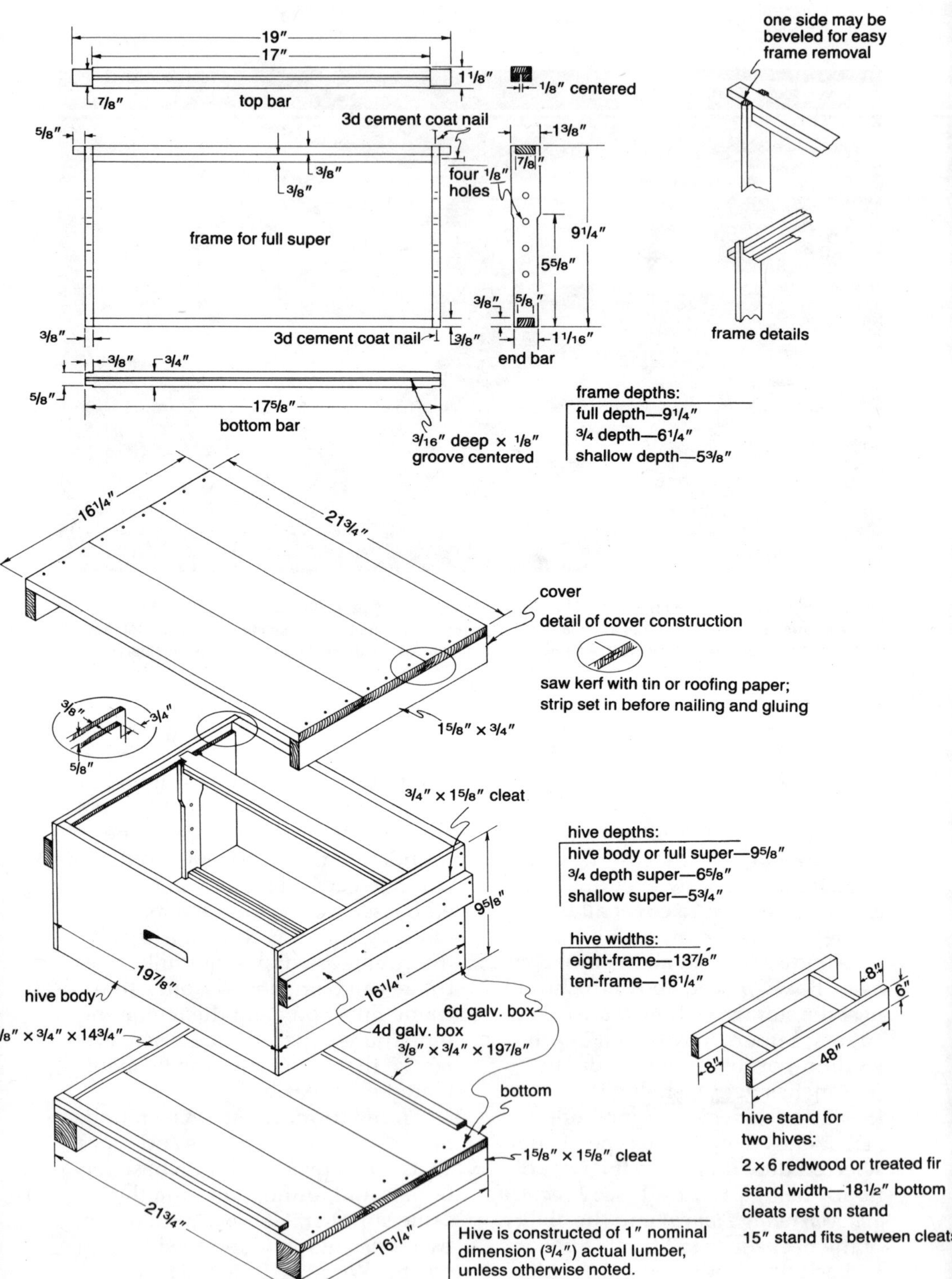

Fig. 3 Dimensions of standard beehives and frames.

and extracting combs. The foundation is thick and reinforced with vertical wires or a plastic midrib. The vertically wired foundation must be supported by additional horizontal wires to avoid sagging under the weight of brood and stored food. Beekeepers can avoid the use of wires, embedding tools, and other problems by using a plastic foundation, which snaps into place, or plastic combs, which combine frame and foundation.

Wiring jig. Beekeepers can purchase or make a jig for wiring frames and embedding wires in beeswax foundation (fig. 4). Best results require use of eyelets in endbar holes to prevent the wire from cutting into the wood. The jig holds a nailed frame under slight end-to-end pressure while No. 26 tinned wire is threaded through the holes and around small spools. One end of the wire is wrapped around a tack which is driven in. The wire is slipped off the small spools, pulled tightly, wrapped around the other tack, and driven tightly. Then it is snipped, usually by bending it back and forth repeatedly.

A jig can be modified to hold an insert that fills the space in the middle of the frame. When a wired frame is placed over a sheet of foundation lying on the insert, the insert presses the foundation up against the wires. An electric train or doorbell transformer should be used to supply the electricity needed to heat the wires and melt them into the wax.

Edible foundation. Beginning beekeepers who do not wish to invest in extracting equipment can produce, without further processing, a class of honey in which portions of comb will be eaten.

Special types of lighter weight comb foundation are used to produce cut comb, chunk, or comb honey sections (see *Producing and Marketing Honey*). Since the foundation is intended to be eaten, it is pressed very thin. The beekeeper should remember that replacing the foundation each time the combs are filled increases the cost of producing comb honey. Recovering the cost of the foundation when the honey is sold is justifiable.

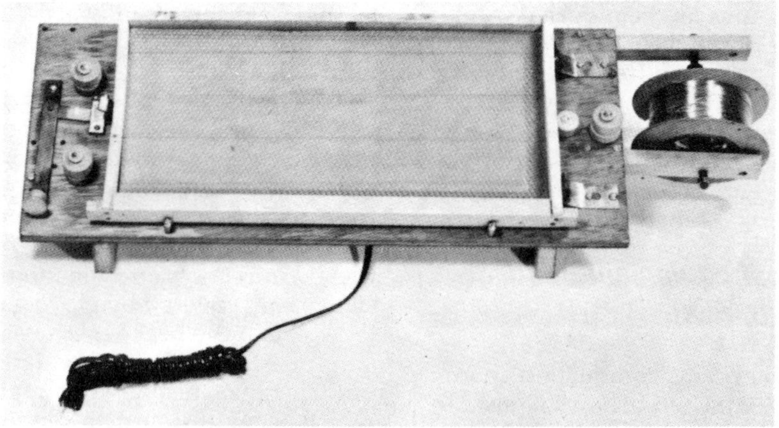

Fig. 4 A wiring jig is used to fasten taut horizontal wires to frames and to embed the wires into sheets of beeswax foundation.

Paint. Hives should be protected from the elements with an ample coating of a good grade outdoor oil, latex, or aluminum-based paint. Paint all surfaces of bottom boards, but only the outside surfaces of boxes and covers. A hive stand (fig. 3) will prolong by many years the useful life of bottom boards. Painting hives different pastel colors significantly reduces the tendency of bees to drift from one colony to another.

Queen excluder. This metal or plastic screen is designed with 0.163-inch spaces to prevent passage of a queen (and drones), while allowing worker bees to pass through (fig. 2*A*). Usually, a queen excluder is placed between the upper hive body and lower super to keep the queen out of the honey storage area. Queen excluders often become clogged with wax, propolis, or drones that can interfere significantly with passage of worker bees to and from the honey storage area.

L-shaped metal rabbets. These strips of lightweight metal are fastened along the edges upon which the frames rest in each box. The metal surface protects the wooden rabbets from damage by the hive tool when frames are being pried from the hive or when propolis and wax are being scraped from the ledges.

Honey extractor. Honey extracting equipment and procedures are described in detail in the section on *Producing and Marketing Honey*.

Personal equipment (fig. 2 D, F, H)

Smoker. This combustion chamber with attached bellows is used to smoke the colony. Smoke puffed into a hive interferes with the chemical alarm system used by bees to alert each other of foreign intruders. Judicious use of cool smoke generated from smoldering burlap sacking, wood chips, or other slowly combustible materials, enables beekeepers to examine their bees with little chance of being stung. Oversmoking bees can cause as much agitation as not smoking them at all.

Hive tool. This tool is specifically designed for prying apart boxes, loosening frames, scraping excess wax, and so forth. The hive tool is essential when manipulating hives because bees collect propolis and use it to seal all cracks between the top box and cover, adjoining boxes, and frames and boxes.

Bee veil. Wire veils commonly are used because they do not blow against the face. Meshed tulle veiling is available in various styles, also. Veils are worn over wide-rimmed helmets and usually have strings to keep them tied to the body. Zipper styles are available for use with matching coveralls.

Gloves. Experienced beekeepers seldom use gloves because their use tends to make handling of frames awkward. However, a pair of kid leather beekeeper's gloves definitely is handy for an amateur working with agitated bees.

Coveralls. White, full-length, zippered cotton coveralls are used by many beekeepers to keep propolis, honey, wax, and smoker exudates from soiling their regular clothes. Additionally, the long pants are tucked into boots and the long sleeves are covered by glove gauntlets to keep the beekeeper nearly stingproof. Avoid wearing dark-colored or fuzzy, heavy-woven fabrics, which seem to stimulate stinging.

Managing Bees

Conscientious beekeepers examine their colonies every 10 days or so from the period of rapid spring population buildup until the beginning of the honey flow, again after each honey flow, and when preparing for winter. Attention must be focused on different concerns during the year, but the basic procedures are the same. The smoker should be lighted and burning well, bee veil and other protective apparel should be in place, and hive tool should be in hand before approaching the hive. It is best not to stand in front of the hive while "working the bees," because it will obstruct returning foragers.

Colony examinations

Smoke the entrance and any holes to the outside, and wait for the smoke to move through the hive's ventilation system. With the hive tool crack the cover and puff a little smoke beneath it. Remove the cover and place it on the ground, to act as a stand for boxes that may be removed from the hive. Carefully remove the outermost frame from the nearest side of the hive body and check it quickly for the queen. The queen usually is not on this frame, but check anyway before setting it down. Very gently stand the frame on edge on the ground, leaning it against a shaded side of the hive. The space created by the missing frame should be adequate to allow the other frames to be pried loose, examined, and replaced out of position in the empty space, until all frames of interest have been examined. During these procedures, glance at the top bars of the frames in the hives to see whether many bees have lined up. If they have, it is time to use a little puff of smoke to send them back down inside. When the inspection is completed, the displaced frames are returned to their original positions, the removed frame is replaced, and the hive is closed. Examinations conducted on warm, calm, sunny days interfere very little with colony functions and are met with little resistance.

Beekeepers should assess critically the following points when they examine a colony:

(1) **Good queen.** This is based on a solid, good-sized brood pattern and/or presence of eggs. There is no reason to find the queen unless the beekeeper intends to requeen.

(2) **Adequate stores of food.** The bees should have a minimum of four

deep frames (filled on both sides) of honey or sugar syrup and the equivalent of one full frame of pollen available to them. Generally, there is more honey, which is essential for overwintering.

(3) **Freedom from disease.** Brood diseases cause discoloration of larvae, or patchy brood with scattered cappings (see *Bee Diseases*).

(4) **Properly arranged hive.** The brood should be kept in the lower hive bodies with empty combs moved to the proper location to allow upward expansion of the brood nest. There should be adequate space above the brood nest for storage of large volumes of nectar and honey if bees are kept in areas with potentially heavy honey flows.

Spring management

An overwintering colony should consist of enough bees to cover at least five frames. If there are fewer, the beekeeper may suspect problems with the queen, disease, or poor late summer and fall pollen supplies, which can commonly occur in California. There should be at least four frames of honey or syrup and adequate space for brood rearing and storage of nectar. Very weak or queenless colonies may be united with more populous colonies.

PREVENTING SWARMING

Swarming is the natural means of colony reproduction. Colonies with an adequate population size rear new queen cells (Plate I), slim down the laying queen, and eventually about half the bees and the old queen leave to seek a new hive and start a new colony. Swarming usually coincides with relatively good foraging periods and tends to occur from late March to July, with a peak in mid-April in the Davis area. Late summer swarming can also be a problem.

Beekeepers who desire maximum productivity from their bees cannot afford to allow half the bees to fly away with a concurrent break in brood rearing of up to 2 weeks. A number of steps may be taken to reduce the chances of swarming, but routine inspections at 10-day intervals and destruction of every queen cell are the only methods by which swarming reliably can be prevented.

Swarming generally is linked to colony congestion. Congestion can be relieved by:

(1) **Reversing boxes.** Bees tend to move their brood nest up to the top of the hive. When young brood fill most of the top box of the brood chamber, put that box on the bottom and allow the bees to move up through other, less filled boxes.

(Consult a textbook for procedures.)

Following the early spring blooming period, a colony should have six or more frames covered with bees in a ten-frame, one-story hive. In a two-story hive, there should be 12 frames covered with bees with brood on both sides of four to five frames (200 to 400 square inches). Under these conditions, brood rearing will increase rapidly, and the colony will build up to a maximum population for the beginning of the major honey flow.

Only a few days of abundant spring nectar flows are needed to crowd the brood-rearing chamber of a one-story hive with honey and pollen, so a super with frames of empty combs should be added to the hive when the flow begins. Frames with foundation should be provided only when the bees are gathering an abundant supply of nectar. Worker bees reluctantly will draw out a super of foundation placed directly above the brood. Inserting a frame of foundation at the edge of an expanding brood nest usually assures acceptance of the comb.

Brood rearing puts a heavy demand upon the food supply. It is important to keep a close check on all colonies during March, April, and May in northern California and as early as December, January, and February, in southern California south of the Tehachapi Mountains.

(2) **Adding boxes.** Bees will move into new boxes to clean the combs, draw foundation, or manipulate honey, thus relieving congestion in the brood chamber.

(3) **Dividing colonies.** When colonies have eight to ten frames of brood, they can be divided into two colonies. It is best to have a young, mated queen ready for the queenless half. (The Demaree method is similar, but division is uneven and both colonies are kept, one above the other, in the same hive. See text books for a full explanation.)

(4) **Using young queens.** First-year queens are much less apt to swarm, while second-year queens and queens from captured swarms are very likely to leave. Clipped (wings) queens are not able to fly; however, this does not preclude swarming. The queen will walk out of the hive, get lost on the ground, and the swarm is very apt to leave later with a virgin queen, leaving the original colony hopelessly queenless. Despite all advertising to the contrary, no special devices meant to be placed in or on the hives can adequately prevent swarming.

If a colony is preparing to swarm, there will be a number of queen cells on the combs. To requeen, select a well-developed queen cell and remove all others. Also find the old queen and kill her by pinching her head. The new queen will emerge from her cell in a few days and thus requeen the colony (unless it swarms!).

Good spring pastures often prove to be inadequate sources of pollen and nectar in summer. Feeding bees both syrup and pollen is essential in some locations for continued brood rearing, maintaining bee populations, insuring adequate winter stores, and successful overwintering. Commercial beekeepers move their hives hundreds of miles to locate their bees in areas with ample forage. Knowledge of the annual blooming sequence and locations of good forage plants is essential for economically successful moves. It is also important to provide adequate space for incoming nectar. Excess open combs have been shown to increase honey production; inadequate space leads to honey in the brood nest, which hampers brood production and leads to population decline.

Extracted honey may be harvested as soon as the combs are full of honey and are three-fourths capped in regions with low relative humidities. Honey should be fully capped and processed quickly in areas with high relative humidities. Equipment and techniques for handling honey are explained in *Producing and Marketing Honey*.

Summer management

Because spring pastures are often inadequate for summer foraging, enough inspections should be made in summer to ascertain whether there is sufficient natural food in the hives; at least 30 pounds of honey should be in each hive all summer. (When inspecting hives, also check for disease.)

PREVENTING ROBBING

A few bees probably rob some honey from other colonies most of the time, but during nectar dearths robbing can become severe and colonies can be destroyed. Robbing bees usually can be recognized by their "dangling feet" flight at the hive entrance and by their attempts to enter cracks between supers.

Taking the following steps will help minimize robbing:

(1) When examining colonies during dearth periods, do not keep hives open any longer than is absolutely necessary and place frames of honey in a super covered with wet burlap during inspections.

(2) If feeding is necessary, start feeding in late afternoon. When bees are fed in the morning, the excitement can trigger robbing behavior.

(3) Colonies are best protected by a robber screen (fig. 5), which reduces the entrance considerably while allowing adequate ventilation on hot days.

Fall and winter management

If they are going to build up quickly the next spring, colonies should go into winter with large, well-fed populations of young fat bees. In many locations in California there is not enough pollen to sustain adequate brood rearing through August and September and colonies should be fed pollen, pollen supplement, or a substitute. (See *Feeding Bees*.) An ideal colony for wintering contains a vigorous young queen, is disease-free, has 10 or more frames covered with bees, and has an adequate supply of stored food. Queenless colonies, or those very sparsely populated, should be united with stronger, queenright colonies. Colonies with less than 10 frames of bees can be overwintered, but they should be forced into a single hive body beneath a super of honey.

The colony actually begins consuming winter stores after the final major honey flow and continues until the spring flows start in earnest many months later. Even in areas where brood is reared year-round and bees attempt to find food daily, honey often is consumed faster than it is produced. A prudent beekeeper should start the winter with 50 pounds (one completely full, full-depth super) of honey or syrup on the bees. Colonies with this amount of food usually will not require winter feeding and will do well during a rainy spring.

"Robbing," an activity evident in apiaries during the absence of a nectar flow, can lead to the destruction of colonies. (See section on robbing.)

As temperatures cool, entrances can be reduced to a couple of inches, but check during winter that entrances have not become clogged with dead bees. Precautions should be taken to fasten covers to the hives and to turn entrances away from prevailing winter winds.

Winter is the slowest season for bees and beekeepers, although winter feeding of bees is common in California. This is the time to prepare equipment, paint hives, and renew supplies of materials to be used the next season. Once February arrives in California, the new beekeeping season begins.

Fig. 5 A robber screen reduces the size of the entrance, which must be protected against robbing bees, while providing a means for hive ventilation.

Feeding Bees

Honey, pollen, and water comprise the natural diet of honey bees. Honey is their carbohydrate/energy source. Pollen provides protein, fats, vitamins, and minerals to the nurse bees, essential for producing the royal jelly that is fed to the queen, drones, and young larvae. Royal jelly is fed in overabundance to larval queens. Worker bees consume a frame of honey and pollen for each frame of brood they rear. A colony rearing 1,000 new bees a day requires nearly 10 pounds of pollen and nectar a month. Simply to survive, various estimates suggest, a colony requires approximately 100 pounds of honey and 50 pounds of pollen each year. Water, the major component of bee tissues, is required to dilute concentrated foods, maintain humidity in the brood nest, and air-condition the hive on hot days.

Beekeepers must examine their colonies frequently to assess the condition of the food supply. During periods of nectar or pollen dearth, the beekeeper must supply substitutes, if continued brood rearing is desired. Supplemental feeding is also used to build up populations:

(1) to compensate in part for pesticide losses,

(2) for winter,

(3) for pollination of almonds, and

(4) for shaking bulk bees for packages.

Instructions for formulating and dispensing various types of feeds to bees follow.

Care should be taken to feed pollen trapped only from colonies known to be free of chalkbrood and American foulbrood disease. Although bees can be induced to rear brood year-round by continually supplying combs with pollen packed into the cells immediately adjacent to the larvae on the next frame, pollen is usually fed to bees as patties during a season of normal brood production. Pollen pellets can be mixed directly into sugar syrups (after being soaked in water for a few minutes) and mixed until a smooth, heavy paste is formed. The mixture should be allowed to sit for 24 hours before feeding to be sure that it has not become too dry, since pollen absorbs quite a bit of the moisture from the syrup. The mixture should be dry enough to prevent oozing between the frames, but moist enough so that the bees can chew it. The 1/2- to 1 1/2-pound patties are placed in the hive on top of the frames, in direct contact with the cluster of bees.

Supplemental Feeding—Pollen		
Product	**Major components**	**Comments**
Air-dried pollen products	Extremely varied: moisture—25%; protein—6 to 40%.	Acceptable to bees if air-dried, then frozen; stores well frozen; mixed pollens better than single source. Soften pellets with water before feeding as paste or mixing into pollen supplement. Has proven ability to spread chalkbrood if trapped from infected colony; can possibly be contaminated by pesticides.

Supplemental Feeding—Sugar			
Product	**Major components**	**Type of feeder**	**Comments**
Honey			
comb	fructose (38%) + glucose (31%) + sucrose (1%) + other sugars (9%) + water (18%)	None	Requires availability of water to be utilized; proven ability to spread American foulbrood (AFB) when taken from a contaminated hive.
extracted		Same as light sugar syrup	Usually diluted by one-fourth or one-half with water. Ferments rapidly; feed only the amount that will be removed from feeder within 48 hours.
Sugar			
granulated	sucrose	None	1 to 2 lb poured toward back of bottom board only to prevent starvation; requires availability of water to be utilized. Not recommended.
confectioner's (powdered)	sucrose + 3% cornstarch	None	Used for diluting antibiotic mixes; can be made into syrup.
Baker's Drivert (powdered)	sucrose + 4% glucose and fructose	None	Very attractive as dry feed; used for diluting antibiotic mixes; used in dry pollen supplements or substitutes; used in production of queen cage candy.
Syrups			
light	1 sucrose: 1 water	Friction top can (or bottle) through hole in cover best; frame-type feeder with footholds; bottle at entrance least effective	Stimulates oviposition, encourages brood rearing, ensures drawing combs from foundation. Used during nectar dearths or to increase colony populations; used in pollen supplements and substitutes.
heavy	2 sucrose: 1 water or 2½ sucrose: 1 water + 1 tsp cream of tartar/20 lb sugar—boiled	Same as light syrup	Fed in late fall, to be manipulated and stored as winter feed when honey is short.
Type 50	sucrose (38½%) + glucose (19¼%) + fructose (19¼%) + water (23%)—acid inverted	Same as light syrup	Delivered as 77% solids and does not ferment or crystalize out of solution; very attractive bee food full strength or diluted any time of year; used in pollen supplements and substitutes; available only by tank truckloads.

—continued

Supplemental Feeding—Sugar *(continued)*			
Product	**Major components**	**Type of feeder**	**Comments**
Liquidose 50	sucrose ($38\frac{1}{2}$%) + glucose (20%) + fructose (16%) + other sugars ($2\frac{1}{2}$%) + water (23%)—inverts from high fructose corn syrup	Same as light syrup	Delivered as 77% solids and does not ferment; will crystalize out of solution below 80°F; very attractive full strength or diluted; used in pollen supplements and substitutes; available in tank truckloads, drums, or 5-gallon pails.
Nulomoline	50 to 90% inverted sucrose + glucose crystals or micropulverized sucrose + water + invertase	None	Very heavy syrup used in preparing queen cage candy.
queen cage candy	Nulomoline + a little Drivert + a couple drops of glycerine kneaded to proper consistency	None	Let stand overnight and check consistency before use; reconstitute if hard and dry or soft and sticky; stores for long periods in airtight containers.
Corn syrup			
high fructose corn syrup (HFCS)	fructose (26%) + glucose (31%) + other sugars (5%) + water (38%)—acid hydration of cornstarch to glucose; enzyme conversion of glucose to fructose	Same as light syrup	Available in various percent solids; no detrimental effects on free-flying bees; may become major bee food if sugar prices increase significantly.

Supplemental Feeding—Pollen Supplements and Substitutes		
Product	**Major components**	**Comments**
Pollen supplement	pollen (5 to 25%) + pollen substitute	Used similarly to pollen substitutes, but added pollen makes mixture more attractive to bees and may be superior nutritionally.
Pollen substitutes	Dry ingredients: a) brewer's yeast b) Torula yeast c) Torutein-10 d) lactalbumin; or mixtures of these	Maintain brood rearing in colonies during periods of pollen dearths. (Do not stop feeding until pollens become available.) Inclusion of some sort of yeast supplies substantial vitamins.
	Moist ingredients: a) light syrup b) diluted honey c) Type 50 syrup d) Drivert syrup	Can be fed dry, but are consumed better as moist patties on the top bars of frames containing brood. Formulated to consistency of thick paste. Should not harden or run down between combs (partially inverted sugars or honey maintain consistency best); consumed rapidly only when bees are rearing brood in absence of pollen; diluted honey can spread AFB if produced in a contaminated hive.
Products to avoid:		
untoasted soyflour		Destroys bees' digestive enzymes.
salt		Only use ingredients with < 2% salt content.

Maintaining Genetic Stock

Good stock is essential to successful beekeeping. Such stock can be obtained from many commercial queen breeders and can be maintained, if care is used in propagation. However, production of stock better than that currently available is usually beyond the resources of the beekeeper. True bee breeding requires use of artificial insemination and a working knowledge of genetics.

In maintaining specific stock, the beekeeper must decide which characteristics are most important and confine selection to preserving them. The beekeeper also must decide how to compile traits in order to compare colonies and measure the results of selection. Selection for high honey production would automatically select for colony population size and its components, as well as for vigorous worker bee foraging.

Environment and heredity

High per-colony yield of honey is desirable in bee breeding, but it is not a simple genetic characteristic. Yield is influenced not only by management and strength of nectar flows, but it is also influenced by the numbers and activities of bees in the colony. The population, in turn, is determined by the egg-laying rate of the queen, the viability of eggs and larvae, and the longevity of adult bees. In stock maintenance, as in breeding, each of these must be considered. Other sought-after characteristics are gentleness and resistance to disease. Each trait probably depends upon heredity.

Stock maintenance

Critical selecting of parents, providing good queen rearing, and assuring an abundance of mature drones from selected mothers are essential. Careful selection of drone mothers is as important as careful selection of virgin-queen mothers. The same characteristics are essential: solid brood pattern, gentleness, and resistance to disease. If bees are to be used primarily for pollination, pollen-gathering ability may be preferred to honey yield.

Ten or more queen mothers should be used to produce queens; a similar number of drone mothers is required to produce drones. Drones in queen-mother colonies should be kept to a minimum; drone-mother colonies should be given frames drawn from drone foundation to encourage production of an abundance of drones. Queens mate with 10 to 20 drones on a single mating flight, so it is important to have a great surplus of drones to assure that virgins will mate with unrelated or distantly related drones. Mating with close relatives can lead to inviable brood.

Care of queens

Frames should always be handled carefully so that the queen is not injured. (See *Managing Bees*.) Never set the frame with the queen on it outside the colony without protecting it with another frame. Do not examine a colony if it contains a virgin, a newly introduced, or a recently mated queen. Such queens are nervous, and the disturbance caused by opening a hive may cause the workers to harm them.

As a rule, the queen should not be handled. If it is necessary to pick her up, catch her from behind by all four wings with the index finger and thumb. Queens can be seen more easily if marked with a color and, by using a different color each year, the queen's age can be readily determined. Fingernail polish, typewriter correction fluid, or other quick-drying paint may be applied to the top of the thorax between the base of the queen's wings.

Requeening

Requeening occurs when a queen of poor stock or one that is aging must be replaced. Colonies can be requeened at any time during the active season, but they usually are requeened during a minor nectar flow because the new queen is more readily accepted then. Before introducing a new queen, the old queen must be removed and any queen cells must be destroyed.

Queen introduction

Queens may be purchased from bee suppliers and are usually shipped in a cage containing a few attendant bees. The cage has an opening that is plugged with bee food known as "candy." Attendant worker bees use the candy to feed themselves and the queen during shipment.

To introduce a queen from a cage into a hive body, remove the cork covering the opening containing the candy plug. If most of the candy remains, ream a small hole through it with a matchstick or small nail. The entire cage is placed in the hive body just below the top bars of two brood frames by gently pressing the bars together so that they hold the cage in place, with the screen side of the cage exposed between the combs. The bees in the hive will consume the remaining candy, releasing the queen and her attendants. This process will take long enough, however, for the bees in the hive to grow accustomed to the new queen.

Pollinating Crops with Honey Bees

Pollination is essential to most flowering plants for producing fruits and seeds, and in California the honey bee is the most important pollinator of commercial crops (fig. 6). Beekeepers rent their colonies for this purpose and place them in or around crops. Highly developed transportation techniques have been worked out for moving colonies from area to area as they are needed.

Many beekeepers have legal contracts with growers for pollination services. A written contract between beekeeper and grower should:

(1) State times (in relation to bloom) when bees will be moved in and out of fields or orchards.

(2) Assure the beekeeper that no pesticides harmful to bees will be used. (If pesticides are to be used, notice must be given to the beekeeper.)

(3) Assure the beekeeper of reimbursement for extra movement of colonies in and out of the field.

(4) Define population of colonies according to numbers of frames covered with bees and either numbers of frames with brood or numbers of square inches of brood.

(5) Describe distribution of colonies in fields and orchards.

(6) Include agreement that the beekeeper will maintain bees in good condition, provided they are not damaged by pesticides while under contract.

(7) State rental rate to be paid to beekeeper.

(8) State who is responsible for supplying adequate water to bees.

Fig. 6 Worker bee gathering pollen from a plum blossom.

Some of the most important factors affecting pollination follow.

Deploying colonies

The number of colonies of bees used per acre depends on the kind of crop, the population of the colonies, the weather, and the amount of competing bloom in the area of the crop. Two colonies per acre in most crops are enough to insure optimum numbers of bees during the most unfavorable conditions for pollination. For alfalfa seed crops, approximately three colonies per acre are commonly used for long-season crop production; however, alfalfa seed yields in areas of short-season production are maximized when growers rent five to ten colonies per acre. In melon pollination, one colony per acre is often used, but two to three colonies per acre are considered better.

Colony strength

In the mid-1960s representatives from the beekeeping and agricultural crop-producing industries agreed to standardize colony strength for pollination purposes: For almond pollination four frames of bees and a laying queen per colony were accepted as the minimum needed, and for alfalfa seed production the minimum acceptable level was at least nine frames of bees and 600 square inches of brood. Growers and beekeepers realize that larger units accomplish more pollination, and recently new pricing structures have been devised that offer a beekeeper bonuses for supplying colonies above these minimum strengths. Saturation pollination often means semistarvation for the bees, and beekeepers should be ready to employ supplemental feeding when necessary.

Distribution

Because bees tend to work close to their hives in attractive pasturage, hives should be distributed at 1/4-mile intervals throughout the fields. If the field is less than 100 acres or the orchard is less than 40 acres, the colonies may be placed in six to eight groups around the edges of the crop. For long-sided, narrow, rectangular fields, colonies should be grouped along each long side, with heaviest concentrations near the center of the field, to provide the best pollinator distribution.

Plant competition

Honey bees may visit plants other than those to be pollinated, if such plants provide more attractive pollen or nectar or if other fields of the same crop are more attractive due to understocking. To prevent bees from visiting competing flowers, colonies should be moved in after the beginning of bloom, when there is sufficient forage to hold bees in the crop. It may be advisable to destroy cover crops in orchards to reduce competition and pesticide hazard.

Other considerations

Weather. Bees begin to fly when temperatures reach about 55°F. They do not fly in rain, heavy fog, or in wind of more than 15 miles per hour. Temperatures above 100°F reduce flight activity for nectar and pollen, but water collection is increased.

Pesticides. Bees provided for pollination services need maximum protection from damage by pesticides. Loss of a portion or all of the foraging bees reduces the amount of incoming food; decreased brood production results. Pollen demand is reduced and pollination decreases. Economic losses to the grower and to the beekeeper are certainties. Damaged colonies may not really recover for months, even if they survive the initial effects of insecticide poisoning.

Bees placed in orchards or fields sometimes are exposed to pesticides applied directly to the bloom upon which they are supposed to be working. Often, too, an application is made to adjacent crops where a competing crop or weed bloom is attracting bees. Beekeepers can avoid or reduce loss by becoming familiar with area pesticide use practices and placing their bees when they feel it is safe. The grower may have to talk to several pesticide-using neighbors if the beekeeper believes that nearby pesticide use is endangering the crop to be pollinated.

The California Department of Food and Agriculture has formulated regulations to aid bee protection, but maximum protection can be obtained only when beekeepers, growers and their neighbors, pest control advisors, and applicators work closely together. Information concerning comparative toxicities of pesticides to honey bees, residual effects, best formulations, and proper timing of applications is available from the University of California Cooperative Extension.

Moving hives

There are very few places in California where colonies can be maintained year-round without encountering prolonged periods of lack of forage. Therefore, in California many migratory beekeepers have developed efficient methods for handling and moving large numbers of colonies. Generally, bees are moved during the night with little or no special precautions taken to confine bees to their hives.

Beekeepers who must spend time loading and unloading hives usually modify their equipment to ease the strain on their backs. Beekeepers with small numbers of hives may use powerlift tailgates or small booms on small flatbed or pickup trucks. Many larger operators have large mechanical or hydraulic booms capable of lifting two hives at a time (Plate I). Booms that can be leveled hydraulically or that are hinged along the boom add flexibility to the system.

In the other major method of moving large groups of hives, four or six hives are fastened to a pallet that also serves as a bottom board for the hives. Pallets of hives can be stacked two-high and loaded with a forklift onto a large flatbed truck. Forklifts are used for distributing hives in and around fields for pollination services. Small forklifts are hauled on trailers. Large forklifts, modified from four-wheel drive trucks, can handle four pallets at a time and can be towed by flatbed trucks. Diesel equipment is used when it can be afforded.

Producing and Marketing Honey

The 1967 Agricultural Code of California defines honey as: "...the nectar of floral exudations of plants gathered and stored in the comb by honey bees. It is levorotatory, contains not more than twenty-five one hundredths (0.25) of 1 percent ash, not more than eight (8) percent of sucrose, its specific gravity is not less than 1.412, its weight not less than eleven (11) pounds, twelve (12) ounces per standard gallon of 231 cubic inches at sixty-eight degrees Fahrenheit."

Honey is composed largely of two simple sugars (glucose and fructose) and enzymes, vitamins, minerals, and substances producing characteristic flavors. California honeys average about 17 percent water. Most honey sold in California is one of three types: extracted (liquid), comb section, or crystallized (creamed or spun).

To be sold in California, honey must meet certain standards as defined in the Agricultural Code (obtainable from California Department of Agriculture, Bureau of Fruit and Vegetable Standardization, Sacramento, CA 95814). Color is important in determining the market value of honey—lighter colors usually bring higher prices. Color varies from nearly colorless through shades of yellows, amber, and brown with greenish tinges, to deep red (Plate I). Honey from the same floral sources may vary in color, and variation in color may result from overheating in processing; for example, honeys darken if heated too much or too long. Color is measured by the Pfund grader or the USDA Color Comparator. These graders tell beekeepers the commercial color classification of honey: water white, extra white, white, extra light amber, light amber, amber, or dark.

Honey is classified according to floral source, method of production, and USDA grades. The two most popular floral honeys produced in California are sage and orange honey. Other major floral sources are: cotton, lima beans, alfalfa, yellow starthistle, wild buckwheat (of the genus *Eriogonum*), manzanita, eucalyptus, and bluecurls. In recent years safflower honey has been produced in quantity.

Marketing

Beekeepers can sell their honey from their homes, from roadside stands on their property, directly to a customer, to a wholesaler in 5-gallon containers or 55-gallon drums, or through a co-op. Cooperative marketing offers certain advantages to beekeepers because a cooperative can control a certain proportion of the total crop and thus increase members' chances for fair prices.

The U.S. Price Support Program (through 1989) includes a minimum base price for honey at the wholesale level. Support prices vary from year to year. For information on this program, inquire at the county offices of the Agricultural Stabilization and Conservation Service (ASCS). The state marketing order for honey is administered by the California Director of Food and Agriculture through the California Honey Advisory Board whose chief function is promoting the use of honey.

Small-scale harvesting

When combs in a super have been filled and capped (two-thirds of cells capped is adequate in a low-humidity environment), the beekeeper may remove a comb or super(s) of combs for harvesting. Be sure to leave the bees one full super of honey and stored pollen for winter feed.

Bees may be removed from each frame by shaking and brushing the bees (with a special bee brush) back into the hive. Frames should be placed in an empty super, sitting on a pallet to keep the combs clean, and covered with a damp cloth to exclude robbing bees.

When there are too many combs of honey to handle individually, bees will remove themselves from supers of fully capped honey in a day or two through bee escape (one-way exit) boards (fig. 7). Place the honey supers above the escape board and smoke the bees gently to get them moving. Two precautions:

(1) Be sure all cracks and holes are sealed or the honey will be robbed.

(2) Be sure outdoor temperatures will stay below 100°F or unattended combs may melt in the hive.

Beekeepers who wish to remove large amounts of honey rapidly should refer to the section on large-scale harvesting.

Preparing honey for storage. Beekeepers without access to extracting equipment can still prepare their honey in manners attractive to consumers. With thin foundations, combs can be cut (called "cut comb honey") easily with a warm, sharp knife into shapes that can be placed into special honey cartons or clear plastic bags after draining on each

Fig. 7 ***Left:*** **Top and bottom of bee escape board; bees enter round hole in bee escape and exit through narrow ends.** ***Right:*** **Top and bottom view of acid board, showing burlap lining on bottom section.**

side for 24 hours. These pieces of cut comb or comb sections (that are produced in special square or round compartments placed in the hives—see the section on honey products below) and whole frames of honey can be held without granulation in a deep freezer for long periods. Pieces of cut comb can be placed in bottles and surrounded with extracted honey; these are called "chunk honey" packs.

Beekeepers who wish to extract liquid honey from combs should use an extractor (fig. 8). Other methods involving chopping, squeezing, or heating combs to the melting point of beeswax usually are considered too messy and wasteful to be worthwhile. Honey produced by 20 colonies usually can be handled by a two- to four-frame extractor operated by hand or by a small electric motor.

Before centrifugation in the extractor, cappings must be shaved from the combs with a heated knife. Thermostatically controlled electric knives are available. The cappings, half wax and half honey, may be bottled and consumed as is (reported useful by some people for hay fever relief) or rendered to save the wax.

Honey flowing from the extractor should be strained first through a coarse mesh, and then through a fine mesh (such as a clean nylon stocking) to remove visible particles of wax, propolis, or other matter. For best appearance, the honey should flow along a flat surface leading to

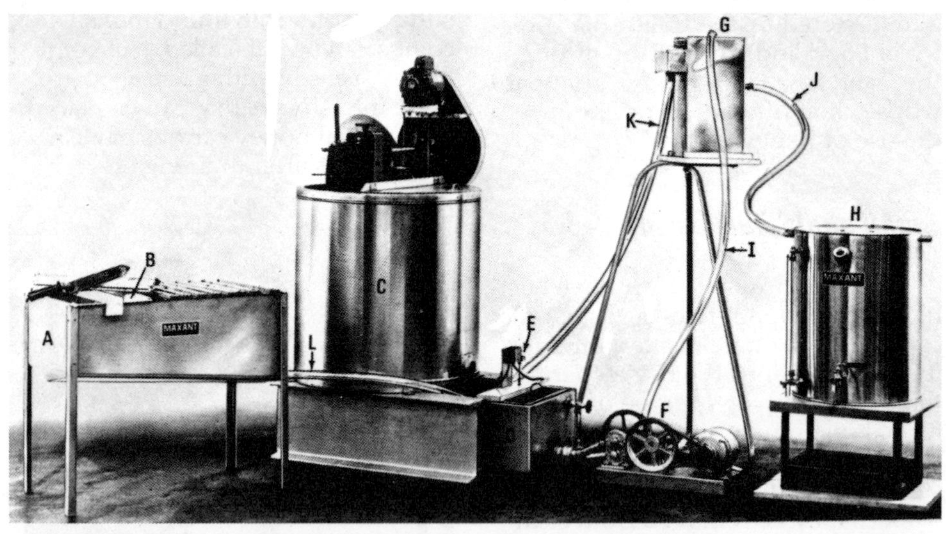

Fig. 8 Honey extracting equipment for a small- to moderate-sized beekeeping operation. *(A)* uncapping tray: *(B)* uncapping basket; *(C)* extractor; *(D)* clarifier; *(E)* float switch; *(F)* pump; *(G)* filter unit; *(H)* bottling tank; *(I)* plastic pipe-pump to filter; *(J)* filtered honey to bottling unit; *(K)* overflow from filter, and *(L)* pipe from tray to clarifer.

the bottom of the containers. Honey that flows as a stream or drips into itself incorporates air and appears somewhat cloudy. Honey handled properly has a clean appearance, full flavor, and all its nutrients intact.

Rendering beeswax. For small operations, a solar wax melter is the simplest and most economical way of rendering wax from well drained cappings and chunks of comb that accumulate over time.

Large-scale harvesting

Most beekeepers in California who remove large quantities of honey use "acid boards," named for carbolic acid which no longer may be used for that purpose. Now, benzaldehyde (oil of bitter almonds) or diluted butyric anhydride is used to moisten the acid board (fig. 7) that is placed directly on top of the super to be emptied. Benzaldehyde is said to work better at lower temperatures (60° to 80°F); butyric anhydride is better on hot days (80° to 100°F). Five minutes should clear the super. Stupified bees signal overdosing; residual bees indicate under-dosing. Be sure to follow label directions and avoid contacting the chemicals or allowing them to contaminate the honey.

A few beekeepers prefer to use bee blowers to empty the supers of bees. A large volume of low velocity air is used to blow bees down a chute onto the ground by the hive entrance. Very few bees take flight, become injured or angered, and fully capped supers are cleared very quickly.

Extracting honey. Supers of honey frequently are stored in a "hot room" in the warehouse before being extracted. Thick, western honeys flow much better when extracted at temperatures between 90° and 100°F. Nearly all beekeepers use mechanical uncappers of one type or another. Uncapped frames are moved by chains to the vicinity of the extractors. Radial extractors, which spin in a horizontal or vertical plane and hold 100 to 350 frames, are most common.

Proceeding from this point in the process, "variety" best describes how the cappings and extracted honey are handled. With one method, all the wax and honey are combined and pumped through a heater and honey separator, which spins the honey away from the wax. Knives in the spinner shave the wax into bits that fall into a barrel.

In many operations, the cappings are routed one way by augers, conveyor belts, or gravity, while the honey is strained and pumped through a second system into settling tanks. The honey is left in the tanks a day or two to allow wax and bubbles to rise to the surface; then it is drawn off from the bottom of the settling tanks into drums, cans, or bottles. Honey stores best at temperatures around 70°F. All types, except sage, will granulate in containers over time, especially at temperatures around 55°F.

Cappings contain valuable beeswax, so various methods are used to separate the wax from the honey and slumgum. One method is to uncap the combs directly into a cappings melter, a water-jacketed tank with heating coils inside and sometimes heating coils in the cover. The object is to melt the wax at about the same speed it is being uncapped. The wax floats above the honey and slumgum, and the levels of honey and wax are adjusted by opening and closing taps leading from the melter.

Most of the honey can be removed from cappings by a cappings spinner, which is similar to a radial extractor but contains a wire-mesh basket. Cappings are added gradually, until the wax builds up to the point that it has to be removed. Lining the basket with pieces of nylon fabric eases wax removal.

A number of beekeepers just collect the cappings in barrels and eventually move them into the oven, a thermostatically controlled, large, insulated box built in the warehouse's hot room. The temperature is maintained at about 150°F and the wax and honey separate in the barrels. The beeswax can be ladled off in a relatively pure state.

A few commercial operations have heated wax presses to extract wax that otherwise would adhere to cocoons in brood combs. Area beekeepers bring barrels of slumgum and old wax to these operations periodically for processing.

Commercial beekeepers should use stainless steel equipment to process honey. Honey is acid, even if it does not taste acidic, and can rapidly corrode many metals and unprotected cement floors.

Honey products

Comb honey. This product is produced in basswood square-section boxes sized 4¼ x 4¼ x 1⅞ inches, in rectangular boxes 4 x 5 x 1⅜ inches, or in round plastic rings that are placed in special comb-honey supers (normal frames are not used in these supers). The exposed surfaces of wooden sections should be painted with hot paraffin after they have been positioned in the supers to prevent bees from staining the boxes with propolis. Overlapping honey flows of light and dark honey should be avoided for comb honey production because combs partly filled with dark and light honey are less attractive to consumers.

HONEYDEW HONEY

Honeydew honey, which is not true honey, originates from a sweet liquid excreted by scale insects and plant lice or aphids. In California, one kind of honeydew honey is derived from a scale insect on incense cedar. Another honeydew honey in California is derived from galls on valley oaks. These galls secrete a sugary material on their exterior walls that is collected by honey bees and stored in the same manner as honey. Major honeydew honey crops have been recorded periodically from valley oaks in the foothills on the west side of the Sacramento Valley for 50 years. Germany is the major market for this hive product.

Comb honey production is difficult because the nectar flow has to be constant and abundant to produce combs of good quality, and because colonies have to be crowded to the swarming point before bees will go into the comb-honey boxes. For best results, there should be enough bees in the colony to fill at least a hive body and a full-depth super, but all should be shaken into a single box; a super of empty comb-honey section boxes should then be added so that the bees can deposit honey in the boxes. After the bees have started to build comb in the super containing comb-honey boxes, a second comb-honey super can be put on top.

When the first combs are about half filled with honey, the two supers should be reversed to finish the first combs (this usually increases honey production). A third comb-honey super may be placed on top, if the honey flow is extremely good and the bees need more space. This can be repeated as long as the nectar flow continues and as often as is needed to give the bees more room in which to work without providing more sections than the nectar flow warrants. (If queen cells are present in such a colony, they should be removed to prevent swarming; the colony that continues to produce queen cells may be indicating that it is attempting to supersede the old queen. Therefore, it may be necessary to kill the old queen and introduce a new one.)

Supers full of comb-honey boxes should be removed as soon as the sections are completely filled and the cells are sealed. Near the end of the nectar flow the comb-honey supers should be removed to permit the colony to store enough honey in the hive for winter. Comb section boxes full of honey should be separated into groups by weight and appearance, and the weight and grade of each section must be listed on its label. The boxes should then be individually wrapped in plastic wrappers and sealed against dust and insects (plastic wrappings are available from bee suppliers). Comb honey is graded according to weight, condition of cell caps, cleanliness, and fullness of combs. Federal standards divide the grades into U.S. Fancy, U.S. No. 1, and U.S. No. 2. Descriptions of these grades are available from the United States Department of Agriculture.

Creamed or spun honey. The objective in producing creamed or spun honey is to induce honey to granulate into such fine crystals that they are undetectable when eaten. Light-colored honeys appear even lighter when processed by this method, so removal of all particulate contamination is mandatory.

For best results, adjust moisture level of honey to 17 or 17.5 percent. Heat honey in a water bath to 140°F to dissolve any natural sugar crystals and to destroy naturally occurring yeasts. Pour the honey through a fine-mesh strainer (equivalent to a nylon stocking's mesh) and cool quickly to 80°F to avoid darkening the honey. Add 10 percent starter, which is good creamed honey, and stir to blend evenly. Try to avoid incorporating air into the mixture because bubbles will cause craters on the surface of the creamed honey later. Pour into containers, cover, and let stand at 55°F. In a week the honey will become very firm. To soften, place at 80°F. To store for a length of time, refrigerate or freeze creamed honey.

Commercial Queen Rearing

Rearing queens

Rearing queen bees on a commercial scale is centered primarily in the Sacramento Valley, which produces approximately 600,000 queens and 900,000 pounds of packaged bees annually for U.S., Canadian, and foreign markets. Skilled beekeepers with many years of experience rear queens, but the following common commercial procedures can be modified to suit smaller operations.

Queens are produced by placing (grafting) 1-day-old larvae into queen cell cups made of beeswax or plastic (fig. 9), and then putting the cups into a functionally or literally queenless cell-builder colony. Young worker bees will feed royal jelly to the larvae until they become pupae, thus assuring that they will develop into queens, not workers.

Almost any colony can be induced to build queen cells, but the quality of queens depends on the care the developing larvae receive. To produce good queens, a colony must have an abundance of nurse bees, pollen, and honey or sugar syrup. Nurse bees can be provided by making sure the colony has a comb of emerging brood a week before grafted cells are given to it. An ample supply of royal jelly can be assured by removing much of the current young brood a few hours before grafting. A day or two before grafting, a comb with pollen should be placed in the colony next to the space to be occupied by grafted cells. Sugar syrup can be fed in a Boardman feeder or by inverting a friction top pail with several small holes in the lid over the occupied frames and inside an empty hive body.

To start this procedure, up to 15 empty cell cups are attached to wooden bars approximately 17¼ inches in length, which fit into 3/16-inch slots cut in the interior sides

Fig. 9 Grafting larvae from comb into queen-cell cups. Good lighting, temperature, and humidity control are required.

of end bars of a standard frame. The cell cups on the bars then are taken to a well lit room with environmental conditions suitable for the grafting. Room temperature should be at least 75°F and the humidity around 50 percent to prevent larvae and royal jelly from drying out. Bright light is important to locating and removing larvae from the combs. A supply of royal jelly (diluted with an equal volume of warm water and stirred until it has an even consistency) and a comb of day-old larvae should be available. A small drop of diluted jelly is placed in the bottom of each cell cup and a larva is placed on the drop. Be sure not to roll the larva over. This operation is repeated until each cell cup contains a larva. The diluted royal jelly keeps the larva moist and well fed. Up to 45 grafted cell cups are placed in the cell-builder colony, which should be well supplied with sugar syrup or honey, and pollen.

After 9 days in the cell-builder colony, the cell cups, containing capped queen pupae (fig. 10), are placed in an incubator (a modified chicken-egg incubator will do) kept at about 91° to 93°F and a relative humidity of about 50 percent. Ten days after grafting, the queen cell is placed in a small mating hive (nucleus or "nuc") containing at least a quarter pound of bees. The bees will take care of the cell until the queen emerges. About 7 days after emerging, the queen flies out and mates with at least ten drones. When the queen mates (in flight), semen from the drones is deposited in her oviducts. The sperm migrate into the spermatheca during the next 15 hours and remain there until used or until the queen lives out her life. The movement of the sperm from the oviducts to the spermatheca is slowed by temperatures below 80°F. Therefore, it is important that the nuclei or colonies be strong enough to maintain a temperature at least this high in the brood area after the queen returns from her mating flight. Newly mated queens begin to lay eggs in 3 or 4 days. The beekeeper then removes the queen from the nucleus and cages her for shipment to a customer.

Fig. 10 Capped queen cells are kept in an incubator overnight for protection and ease of handling.

During shipment to a customer, the queen can be in a cage, along with a few attendant bees and a supply of queen "candy" for food. In a shipment of a queen and enough bees for a colony, the queen is sent alone in an empty cage placed in the package full of bees. The package contains a can of sugar syrup. The bees will feed the queen through the wires that enclose her cage.

Packaged bee production

Packaging bees is an enterprise for experienced beekeepers, but the procedures described below can be modified for hobbyist operations.

Colonies expected to produce bulk bees should be fed well in late summer, fall, and early spring. This stimulates longer and better brood production, which results in large bee populations that can take maximum advantage of early spring blooms. With such strong colonies a beekeeper who places bees in almonds in February and near wildflowers in March can maintain strong colony populations even though many pounds of bees may be removed for packaging.

Removal of bees is referred to as "shaking." To shake bees, one or more frames are removed from each hive and shaken with a downward jerk over a funnel that empties into a "shaker box." However, beekeepers now smoke the bees up through a queen excluder into a screen-topped box with vertical panels on which the bees cluster. The clusters are jolted down through the funnel into the shaker box. The filled box is then carried to where the packages are weighed; here the bees are shaken from the shaker box into a weighing vessel that transfers the bees into packages built for bee shipping (fig. 11). Two to 4 pounds of bees may be shipped in the packages, each of which contains a can of sugar syrup (except when shipped by air). The packages are nailed together with laths in groups of five; this makes them easier to handle and provides space for adequate ventilation when they are stacked. A 2-pound package is considered the minimum for starting a new colony; a 4-pound package with two queens may be divided into two potential colonies.

Fig. 11 Bulk bees are weighed and poured into wire-screen boxes containing caged queens. With a can of sugar syrup in the hole, the packaged bees are sent to beekeepers to populate empty hives.

Other Enterprises

Beeswax

Beeswax, used mainly for comb foundation, is an important commodity to the cosmetics industry, candlemaking, and other manufacturing. California produces approximately 600,000 pounds annually—about 10 percent of the total produced in the U.S. (The U.S. imports more beeswax than it produces, a fact that might influence a beekeeper's goals.)

Beeswax, secreted by wax glands located on the underside of the bee's abdomen, is used by worker bees to construct combs. To achieve maximum beeswax production, large quantities of honey or sugar syrup must be present in the colony because bees must consume about 8 pounds of honey to produce 1 pound of wax. In preparation for wax production, bees gorge themselves with honey and hang in chains called "festoons." In about 24 hours wax secretion begins.

For small operations a solar wax melter is the simplest and most economical way of rendering wax. (See UC publications listed in *References*.) Commercial beekeepers use steam-heated tinned copper, galvanized iron, aluminum, or stainless-steel containers to render beeswax. A few operations use a heated wax press to extract wax that otherwise would adhere to the cocoons in brood combs.

Most beekeeping supply dealers buy raw wax or will exchange it for foundation. Clean light wax from comb cappings yields the highest market price.

Royal jelly

Royal jelly, used to prime queen cell cups, raise bee brood in the laboratory, and by some individuals for medicinal or cosmetic purposes, can be produced year-round in some areas in California. Good production requires the strongest colony that can be produced, adequate nectar and pollen flow, and feeding of colonies when natural sources of food are lacking. A queenright colony is usually used in royal jelly production, because queenless colonies require more labor to keep colony populations high. Any one of the following three types of hive setups is effective.

(1) Customarily, a one-story hive with nine frames is divided so that there are five frames for normal

brood rearing and four frames for royal jelly production. To create a division, a sheet of window-screen wire and a sheet of plastic are fastened to a strip of 1/8-inch by 1-inch wood so that the wood is between them. The wooden strip, screen, and plastic each should be long enough to extend the complete length between the front and back walls of the hive body. The screen should be long enough to reach the bottom board, and the plastic should be long enough to cover the top bars of five adjacent frames. The long side of the screen and the plastic should be fastened to the wood.

After the frames are moved to the appropriate positions and the worker bees have an opportunity to reposition themselves, the screen is inserted between the fifth and sixth frames, and the plastic sheet is laid over the five frames upon which the queen is allowed to roam. This keeps her from moving to the adjacent group of four frames where royal jelly is to be produced. The four frames on the royal jelly side consist of two outer frames of honey and pollen, one inner frame of open (young, uncapped) brood, and another inner frame containing a bar upon which have been grafted 15 to 20 queen cells.

The nurse bees of the colony will be stimulated to secrete royal jelly to feed the larvae in the queen cells as well as the uncapped brood in the adjacent frame. The frame holding the grafted cells should be removed on the fourth day after the grafts have been placed in position. The cells are then trimmed down to the level of the royal jelly they contain and the larvae removed. An aspirator is used to remove the royal jelly from the cells. The jelly is packaged in 1-pound, airtight Opalite ointment jars and refrigerated at 40°F. Usually, jelly is shipped to the buyer as soon as possible after harvest and by the fastest transportation possible.

(2) Another setup consists of a two-story hive, with the previously described one-story hive used as the upper story (or super) above a queen excluder; each story has its own queen. Under these conditions additional nurse bees can move up from the bottom hive to help rear the cells.

(3) A two-story hive has a queen and a colony of bees in the lower hive body, separated from the super by a queen excluder. An open wooden rectangle (made of 1-inch stock) with outside dimensions identical to the hive body is placed above the queen excluder and a super is placed on it. This holds the bottom of the frames in the super an extra inch above the top bars of the frames in the hive body, and thus reduces the probability of crushing bees when frames are placed in the super. Nine frames are used in the super. The center frame has a bar with 15 to 20 grafted cells on it for jelly production. On each side of this frame is a frame of open brood and nurse bees. The other six frames are for honey and pollen.

Regardless of the setup used, the colony must be managed in certain ways for maximum production:

- Frames of originally open brood on either side of the frame holding the bar of grafted cells must be replaced with new frames of open brood at least every 14 days, or after every fourth graft. To produce open brood, frames with empty cells are placed in the brood nest area of the queenright portion of the hive.

- A residue of jelly is left in each cell when the jelly is harvested. A drop of royal jelly diluted by one-half with water is placed on the residue of jelly in the cell before grafting because the larvae float off the grafting tool more easily onto this diluted drop.

- The cells producing royal jelly should be grafted again as soon as possible after the previous jelly has been harvested.
- Larvae are grafted from frames of brood borrowed from the queenright section of the royal jelly producing colonies.
- The queen sometimes cannot produce enough nurse bees, but this can be corrected by requeening with a more prolific queen. More bees can be added by placing combs of emerging bees from other colonies into weak colonies.
- Each colony should be harvested every fourth day. One pound of royal jelly per 200 colonies is usually produced under the best conditions, which are the same as those for honey production.

For best quality, royal jelly should be shipped as quickly as possible after production.

Harvesting pollen

Bee-collected pollen has a number of uses besides providing nutrients to bees. Pollen has become a permanent supplement in many peoples' diets. Extracts of pollens are important in desensitization to hay fever.

Pollen-collecting activities of bees can be monitored to determine efficiencies of colonies used for pollination services. It should be noted, however, that any pollens that contain toxicants should be avoided. California buckeye pollen is toxic to bees as are pollens that are contaminated with pesticide residues.

Pollen is carried to the hive in ball-like pellets on the hind legs of foraging bees. A portion of the pollen pellets may be harvested by placing a pollen trap on a hive. Generally, the hive is rearranged so that the pollen trap serves as the entrance; it may be located at the bottom, middle, or top of the hive. After the bees become accustomed to the new entrance, a sieve of two layers of 5-mesh hardware cloth, separated by 1/4 inch and with holes offset, or a perforated (3/16-inch holes on staggered centers) metal plate, is positioned in the trap. The sieve knocks off some of the pellets that fall into a collection tray. The tray should be equipped in some manner to reduce the moisture content of incoming pellets (often a screen or cloth bottom) and should have an 8-mesh, thin wire, hardware cloth cover to prevent bees from entering and recovering the pollen. Pollen traps may be purchased from suppliers of beekeeping equipment, but many individuals prefer to design their own.

Pollen traps should be used only during heavy pollen flows. At the hive, bees will increase emphasis on pollen collecting to make up for what is lost, but this is possible only when there is spare pollen within flight range.

Pollen pellets should be removed from the traps every 2 or 3 days. The pellets should be spread out and allowed to air dry for 24 hours to reduce the potential for mold growth. Then the pollen can be placed in containers and stored at 0 °F. Freezing increases the storage life of the product well over a year and eliminates all stages of stored product insects, which can become a serious problem. Frozen pollen becomes moist and crumbly when thawed and should be used immediately upon removal from the freezer.

An alternative method for storing pollen, developed at the University of Guelph, Ontario, Canada, is to mix the pollen with half its weight of fine granulated sugar. Pack the mixture to within a 1/2-inch layer of the top of the storage container, cover the mixture with a 1/2-inch layer of sugar, and seal airtight. Pollen stored mixed with sugar remains soft and moist and can be formed into cakes and fed directly to bees or mixed into a pollen supplement.

Sources of Nectar and Pollen

Hundreds of species of California plants yield pollen or nectar, but the most important plants for honey production are alfalfa, oranges, cotton, lima beans, sages (black, sonoma, white, and white leaf), yellow starthistle (fig. 12), wild buckwheats, manzanita, eucalyptus, and bluecurls. Extensive use of herbicides to control weeds has decidedly reduced bee pasturage in California. Beekeepers should actively encourage the use of plants beneficial to bees for plantings along roadsides and other rights of way, revegetation of disturbed lands, and ornamental plantings.

Alfalfa, oranges, cotton, corn, and beans present a hazard for bees because of the pesticides used on them.

California buckeye, *Aesculus californica* (Spach) Nutt., deserves a special note of caution because of its toxic nature to bees and its wide distribution and abundance. The tree is found throughout the foothills of cismontane (see *Glossary*) California from Siskiyou and Shasta counties to Kern County and northern Los Angeles County below 4,000 feet. It blooms in May and June and is very attractive to honey bees, but when buckeye pollen becomes predominant in the diet of larval bees, malformed nonfunctional adults result. (See complete discussion in *Other Disorders*.)

Fig. 12 Yellow starthistle, a highly prized honey plant of northern California.

Wild Plant Sources of Both Nectar and Pollen*

Plant	Where found	Time of bloom	Color of honey
Bluecurls (*Trichostema lanceolatum* Benth.)	Dry open fields below 3,500 feet; most of cismontane California.	August to October	White
Buckwheats, wild (*Eriogonum* spp.)	Throughout California.	April to November	Light amber
Buttonbush (*Cephalanthus occidentalis* L.)	Along streams and lakeshores, below 3,000 feet; throughout Central Valley and adjacent foothills.	June to September	White to light
Clover, Brewer (*Trifolium breweri* Wats.)	Wooded slopes below 6,500 feet; in Sierra Nevada from Madera County north; in Trinity, Siskiyou, and Del Norte counties.	May to August	White
Deervetch: broom; wild alfalfa (*Lotus scoparius* [Nutt.] Ottley)	Dry slopes, often following burns below 5,000 feet; most of cismontane California.	March to August	White
Eucalyptus (*Eucalyptus* spp.)	Central Valley and Coast Ranges south to San Diego County.	December to July	Light amber
Filaree (*Erodium* spp.)	Stock forage plant of open grassy areas, below 3,500 feet; throughout cismontane California.	February to August	Light amber
Goldenweed fleece (*Haplopappus arborescens* [Gray] Hall)	Dry foothills below 4,000 feet (to 9,000 feet in Sierra Nevada); cismontane Sierra from Nevada County to Tulare County; Coast Ranges from Del Norte County to Ventura County.	August to November	Amber
Jackass clover (*Wislizenia refracta* Engelm.)	Alkali plains in San Joaquin Valley. Very limited distribution due to land reclamation.	April to November	Water white

—continued

Wild Plant Sources of Both Nectar and Pollen* *(continued)*

Plant	Where found	Time of bloom	Color of honey
Mesquite (*Prosopis glandulosa* Torr. var. *torreyana* [L. Benson] M. C. Jtn.)	Washes below 3,000 feet; Colorado and Mojave deserts, San Joaquin Valley, and interior valleys from Santa Barbara County to San Diego County.	April to June	Light amber
Mountain misery (*Chamaebatia foliolosa* Benth.)	Open forest, 2,000 to 7,000 feet; Shasta County to Kern County.	May to July	Amber
Mustard (*Brassica* spp.)	Weeds of orchards, open grassy slopes, and waste places; throughout cismontane California. Limited by weed control.	January to May	Light amber
Poison oak (*Rhus diversiloba* T. & G.)	Low places, thickets, and wooded slopes, below 5,000 feet; throughout cismontane California.	April to May	White
Rabbit brush (*Chrysothamnus nauseosus* [Pall.] Britt.)	Dry, open plains, and mountainsides, 2,500 to 9,500 feet. Mostly in transmontane California.	September to October	Light amber
Sage, black (*Salvia mellifera* Greene)	Dry slopes, below 2,000 feet; Coast Ranges from Monterey Bay to San Diego County.	April to July	Water white
Sage, Sonoma [creeping] (*Salvia sonomensis* Greene)	Dry slopes, below 6,500 feet; foothills of Sierra Nevada from Shasta County to Calaveras County, and Coast Ranges from Siskiyou County to Napa County, plus Monterey, San Luis Obispo, and San Diego counties.	May to June	White
Sage, white (*Salvia apiana* Jeps.)	Dry slopes, below 5,000 feet; Santa Barbara County south to San Diego County.	April to July	Water white
Sage, white leaf [purple] (*Salvia leucophylla* Greene)	Dry slopes, below 2,000 feet; Orange County north to Monterey and Kern counties.	May to July	Water white
Sages other than black, Sonoma and white leaf (*Salvia* spp.)	Mountain ranges and foothills throughout California, mostly below 5,000 feet.	Spring and summer	Water white
Spikeweed and tarweed (*Hemizonia* spp.)	Dry, open slopes and grassy fields below 3,000 feet; throughout cismontane California.	April to November	Light amber to amber
Starthistle, yellow (*Centaurea solstitialis* L.)	Widely distributed weed. Limited by extensive weed control program. Was widely spread in Sacramento Valley.	May to October	White to extra light amber
Sumac, laurel (*Rhus laurina* Nutt.)	Dry slopes, below 3,000 feet; cismontane southern California from Santa Barbara County to San Diego County.	June to July	Amber
Toyon (*Heteromeles arbutifolia* M. Roem.)	Brushy slopes and canyons below 4,000 feet; foothills and mountains of cismontane California.	June to July	Amber
Wild lilac (*Ceanothus* spp.)	Dry slopes, often rocky or wooded, mostly below 6,000 feet; foothills and mountains of cismontane California.	March to July	White
Willow (*Salix* spp.)	Streambanks, meadows, and wet places; throughout California.	January to July	Amber

*See *California Flora*, Munzand Keck, 1959. Berkeley: University of California Press.

Bee Diseases

Bee diseases are specific to either the brood or adult bees. Brood diseases generally are considered more detrimental to the colony than are adult diseases. No disease of bees affects humans.

Before attempting to diagnose brood diseases, the beekeeper should become familiar with the appearance of healthy brood in all stages (Plate II). In healthy colonies there is regularity in the arrangement of eggs, open larvae, capped brood, and emerging bees. Healthy larvae in open cells are plump, glistening, and pearly white. Brood cappings normally are uniform and raised slightly above the comb surface. Once in place over the larvae, the cappings remain free of visible holes until emerging bees cut their way out of the cells.

Brood diseases

Symptoms of brood disease generally first become noticeable in combs containing mature brood where young bees are emerging or in combs containing more than one cycle of brood. Symptoms also may be found in brood combs of a hive in which the colony has died of a brood disease. Symptoms can be observed readily only after the adult bees have been shaken off the comb.

A comb being examined should be tilted so that direct sunlight illuminates the lower side walls and bottoms of the cells. This makes it possible to see any disease scale that might be present. If no dead brood is found in open or uncapped cells, it is advisable to remove any sunken, discolored, or punctured cappings and examine the cell contents. When dead brood is found, the following should be noted: position of dead brood, age and type of brood affected, color, consistency, odor of dead brood in various stages of decay, and position and tightness of scales.

The consistency of decaying larvae is important in disease diagnosis. Consistency can be determined by stirring the decaying larva with the larger end of a toothpick and slowly withdrawing the adhering mass, observing the texture and noting whether the material can be stretched out into a thick thread. This threadlike property is termed ropiness. Used toothpicks should be burned in the smoker or wrapped in wax paper and sent to the California Department of Food and Agriculture for analysis, if American foulbrood is suspected.

American foulbrood (AFB). This, the most serious larval bee disease in North America, is found in California and at times has made commercial beekeeping unprofitable in some areas. Colonies should be inspected periodically for this disease.

Larvae dead of AFB lie fully outstretched on the lower cell walls (Plate II). Pupae also may be killed and usually die with their "tongues" stretched vertically across the cells. Diseased larvae or pupae gradually change color from white, through butterscotch, to dark coffee brown, and finally dry to become thin scales that adhere tightly to lower cell walls. Decaying brood is slimy and ropy, and a burned odor is often noticeable.

American foulbrood is caused by a bacterium, *Bacillus larvae* White. Spores enter the bee larva in contaminated food, and bacteria resulting from these spores multiply and kill the larva in its cell, usually just after the cell has been capped. The bacteria continue to multiply in the dead tissues and cause decay. As the decaying mass dries, the bacteria transform into highly infectious spores, billions of which may be contained in a single dried scale. Spores of *B. larvae* are extremely resistant to high and low temperatures, to chemical disinfectants, and to the dehydrating action of honey that normally kills bacteria. The spores can remain alive and infectious for decades in honey, in combs, and on used equipment.

The disease is spread within the colony by adult bees whose mouthparts are contaminated from working with nectar or honey containing spores, or by attempts to remove diseased brood. Spores from contaminated mouthparts of nurse bees are incorporated into larval brood food and the larvae become infected. Nectar, honey, and pollen collected from hives in which the bees are infected with AFB are contaminated with spores and may cause disease if given to a healthy colony.

American foulbrood is persistent, and infected colonies soon become unable to rear enough healthy brood to maintain colony populations. As infection progresses, the colony is weakened, honey and combs in a diseased colony become heavily contaminated with spores, and the disease is spread when robbing bees bring contaminated honey back to neighboring colonies. Thus, diseased colonies constitute a serious menace in any area where bees are kept.

To control AFB and prevent its spread, colonies must be inspected regularly and diseased colonies must be destroyed. After the disease has been discovered and abated in an apiary, the remaining colonies should be thoroughly examined again in 30 to 60 days. A beekeeper not experienced in AFB abatement who finds signs of this disease should contact the county agricultural commissioner immediately to obtain assistance.

Drugs may be fed to colonies not showing AFB disease symptoms as an aid in preventing the disease. Terramycin is registered for such use and has appropriate directions on the container label. It is illegal in California to medicate or otherwise maintain an AFB-diseased colony of bees.

European foulbrood (EFB). This disease, which occurs in some parts of California, can seriously cripple a colony. Strong colonies can usually recover.

Larvae dead of EFB are coiled in the cell bottoms or are twisted across the lower cell wall; occasionally, they die in an outstretched position. Diseased larvae first turn yellow, then brown, and finally dry to form dark, irregularly shaped scales that are easily removed from the cells. Decaying larvae have a wet, paste-like consistency; sometimes exhibit a degree of ropiness; dry to form scales closely resembling those of AFB disease, and give off a sour odor. Pupae rarely are affected. It is rare that a larva, dead of EFB, is found with the head upstretched to resemble the tongue mentioned in the previous discussion of American foulbrood.

Sometimes EFB appears suddenly and spreads rapidly through colonies. This is most likely to occur in spring after the first or second brood cycle and during a pollen dearth. At other times, it may spread slowly and do little damage. A good nectar flow seems to hasten recovery. The disease usually subsides by midsummer, but occasionally remains active during summer and fall; or it may seem to disappear and then reappear in fall. Terramycin is registered for use in treating EFB-diseased colonies. Label directions must be followed closely to avoid injury to brood or contamination of honey.

European foulbrood often can be controlled by dequeening the diseased colony for 10 days, which breaks the brood-rearing cycle and gives the bees an opportunity to clean diseased brood from the cells. The colony then can be requeened and more bees added, or it can be united with a stronger colony after the 10-day period. This treatment frequently is also effective against parafoulbrood.

Parafoulbrood. Occasionally found in California, parafoulbrood is comparable to EFB in its effect. Larvae dead with parafoulbrood typically lie twisted across the lower cell walls, although sometimes those lying in a normal coiled position may also be dead. Occasionally, older larvae in sealed cells are killed and lie outstretched. The decaying mass has a reddish brown color, and the texture usually is moist and pasty or gummy; occasionally, there is ropiness. This disease is easily confused with either AFB or EFB and sometimes appears to be a mixture of the two.

Sacbrood. Sacbrood seldom results in more than slight damage to colonies. Larvae killed by the disease lie fully outstretched on the lower cell walls and are usually yellow or brown and darker at the head end. The larval skin remains intact, and the body contents become watery, making it possible to remove diseased larvae intact as a fluid-filled sac. The odor is sour. Scales are brown and wrinkled with turned-up ends, and they are easily removed. Sacbrood is more prevalent in spring but usually clears up with a good nectar flow. Colonies in which the disease persists can be requeened with resistant stock or united with stronger colonies after killing the susceptible queen.

Adult bee diseases

There are two major diseases of adult bees, nosema disease and paralysis. Nosema disease is endemic in many colonies, but its detrimental effects are very subtle. Paralysis occurs sporadically with symptoms very similar to insecticide poisoning.

Nosema. Caused by a microscopic protozoan, *Nosema apis* Zander, this disease seldom causes mortality. The only reliable way to determine whether *Nosema* is present in a colony is to submit a sample of bees for laboratory examination. *Nosema* infections become much more severe during bad weather, when bees are confined to hives. An infected colony may become seriously weakened during the critical population buildup period in spring. Infected bees have their lives shortened as much as 40 percent, even though they appear to forage normally until shortly before death. More importantly, however, infected nurse bees rapidly lose the ability to produce the royal jelly required to feed the queen, the brood, and the drones. Lack of royal jelly leads to population stagnation or decline during periods of anticipated population explosion, even when there is an abundance of spring pollen and nectar sources. These negative effects often are referred to as "spring dwindling." Severe infections can lead to off-season (winter) supersedures that produce drone layers (permanently virgin queens) or queenlessness by the next spring.

Losses from *Nosema* are not always immediately apparent, as it is common for infected bees to die away from the hive. It usually is too late to apply effective control when signs of the disease are seen, so prevention is the only alternative. Fumagillin, sold under various trade names, is registered for preventive treatments. The disease may be held to a minimum by keeping colonies strong and by overwintering them in locations sheltered from wind and open to maximum sunshine. Confinement of bees, pollen shortage, unripened winter feed, chilling, and frequent handling can accelerate buildup of *Nosema*.

Paralysis. Paralysis disease is widely distributed in California. It seldom causes serious damage, except occasionally in southern California, and affected colonies usually recover.

Diseased bees may be seen on top bars or at the colony entrance. Typically, they are weak, they shiver or tremble, and they are unable to fly or walk in a coordinated manner. Frequently their legs are widely sprawled, their wings disconnected, and their bodies hairless, with a dark, greasy appearance. They have a distinct and repulsive odor.

Paralysis is a mildly infectious virus disease transmitted directly from sick to healthy bees. Colonies in which the disease seems to persist should be requeened with a less susceptible stock.

Honey bee parasitic mites

The federal Bee Importation Act of 1922 prohibited further importation of live honey bees into the U.S. The Act was passed in response to concern about a devastating loss of bees in Europe that was associated with the presence of microscopic mites in the thoracic tracheae of the bees. The impact of infestation by the tracheal mite, *Acarapis woodi*, which was detected in the U.S. in 1984, appears to vary, depending upon the stocks of bees and environmental conditions. In California, quarantine regulations are in effect to keep the tracheal mite out of northern California. Check with your county agricultural commissioner about regulations, because they change periodically.

Varroa jacobsoni, a larger external parasitic mite of larval, pupal, and adult honey bees, has eventually made its way from the Far East (originally hosted by *Apis cerana*) to South America. This reddish brown little mite, shaped like a crab with no big pincers, appears to be more destructive to European bees than is *Acarapis woodi*, particularly since it damages brood in addition to feeding on adult bees. Every effort is being made to keep this parasite out of the U.S.

MATERIALS REGISTERED FOR BEE DISEASE CONTROL

Oxytetracycline hydrochloride (American foulbrood control)

Commonly purchased as Pfizer's product, Terramycin, and referred to as TM. A formulation strength of 25 grams of oxytetracycline in 1 pound of mixture is referred to as TM 25. Other concentrations are available.

It is recommended that the purchased formulation be diluted down to TM 5 (1 part TM 25: 4 parts sugar) with powdered (confectionary or drivert) sugar and that 2 level tablespoons be applied carefully along the top bars of combs containing eggs and young larvae in a well established colony. Each treatment provides about 10 days of protection. The antibiotic should be used only when brood is being reared and when there is a chance that foraging bees may be robbing sick, weakened colonies during a nectar dearth (spring, early summer, fall). Medication should be terminated 4 weeks before an anticipated nectar flow to prevent contamination of honey.

Fumagillin (nosema disease control)

Currently available under a variety of trade names. Sold as a powder, it gives reliable results only when administered to the colony in sugar syrup at the concentration recommended on the label. Colonies overwintered in cold areas should be fed 2 gallons of medicated sugar syrup in fall. Colonies overwintered in warmer regions can be fed 1 gallon of medicated heavy sugar syrup in fall and the second gallon of medicated light sugar syrup in January when bees are being stimulated into early brood rearing.

Fumagillin persists for a long time in sugar syrup, so it should not be fed when early nectar flows tend to be included in the harvested honey crop.

If you see mites in your beehives, send samples, in alcohol, to the Supervisor of Apiary Projects, California Department of Food and Agriculture, 1220 N Street, Sacramento, CA 95814. It is critical to find and eliminate mite pests before they cause irreparable damage to the state's beekeeping industry.

Diagnosing diseases

If there appears to be a mixture of symptoms or if symptoms do not seem typical, it is advisable to submit samples to the Supervisor of Apiary Projects. Diagnosis is free.

To take samples of diseased brood, a smear should be made by stirring the cell contents with a clean toothpick. Transfer the smear to a small piece of waxed paper along with the toothpick, fold the paper to prevent contamination, and place in an envelope, along with a letter requesting diagnosis. Samples of scale pried loose from cell walls may be submitted in the same manner.

Samples of adult dead bees should be fresh; dried specimens are of little value. Samples should be selected carefully and made up only of bees that appear to be affected. They should be mailed in a sturdy container that will protect them from being crushed. Live bees also may be sent, provided they are properly caged. Satisfactory diagnosis can be made from a sample of 10 to 20 bees.

Comb samples are difficult to handle and are unnecessary in laboratory diagnosis for AFB determination.

Other Disorders

Healthy colonies do not tolerate the presence of dead brood and will remove it as quickly as possible. Seriously weakened colonies should be strengthened or united with stronger colonies to hasten removal of dead brood from the combs.

Poisoning

The first sign of poisoning usually is the appearance of a large number of dead or dying bees at colony entrances throughout the apiary. A knowledge of local pesticide programs and of blooming plants that are toxic to bees is important in avoiding poisoning.

Pesticide injury. Some pesticides kill larvae in all stages as well as adult bees. Pesticide dusts are particularly hazardous because of their greater tendency to drift and their propensity to stick to bee hairs. If pesticide damage is suspected (fig. 13), the beekeeper should immediately file a Report of Loss with the agricultural commissioner of the county in which the damage occurred so that the loss can be recorded and investigated properly. Beekeepers can request the commissioner's office that they be notified when pesticides they consider particularly dangerous to their colonies are going to be applied in their area — they will be given a 48-hour warning. A publication that includes information on the potential hazards of specific pesticides to bees can be obtained at any farm advisor's office.

Poisonous plants. California plants producing nectar or pollen poisonous to bees are: California buckeye (*Aesculus californica* [Spach] Nutt.), death camas (*Zigadenus veneosus*), cornlily (*Veratrum californicum*), and locoweed (*Astragalus* spp.). Because of its wide distribution, buckeye is the most hazardous to bees.

Symptoms of buckeye poisoning usually appear about a week after bees begin working the blossoms. Many young larvae die, giving the brood pattern an irregular appearance. The queen's egg-laying rate decreases or stops, or she may lay only drone eggs; after a few weeks, an increasing number of eggs fail to hatch or a majority of young larvae die before they are 3 days old. Some adults emerge with crippled wings or malformed legs and bodies. Foraging bees feeding on buckeye blossoms may have dark, shiny bodies and paralysislike symptoms. Affected colonies may be seriously weakened or may die. However, the

Fig. 13 Bees killed by pesticides.

queen may resume normal egg laying, if the colony is moved from the buckeye area.

Honey produced from California buckeye is not poisonous to humans. (Oddly enough, neither is honey produced from poison oak.)

Brood disorders

Chilled or starved brood. Chilling usually occurs in early spring when a severe drop in temperature follows warm weather — this causes the bee cluster in the hive to contract and no longer cover brood in peripheral portions of the brood-rearing area. Chilling also can result from separating brood combs from the main brood-rearing areas or from neglect by worker bees if too many bees are lost to pesticides or other causes. Brood dead from chilling or starvation will be found in a clearly defined area, not scattered among healthy brood. All stages of brood in the affected area will be dead. Dead brood is gray, brown, or black and has a slightly sour odor; it is easily removed from cells. Symptoms usually disappear with warmer weather, with supplemental feeding, or with the beginning of a nectar flow.

Overheated brood. Brood dead from overheating resembles brood dead from chilling or starvation. When such brood is found at the same time that older larvae are observed crawling outside their cells, overheating has occurred. To prevent this, bees should have adequate ventilation, an ample supply of water, and shade.

Dead drone brood in worker cells. If a queen bee lacks sperm, she will lay only drone-producing (unfertilized) eggs. If no queen is present, worker bees sometimes will lay eggs, but these will be unfertilized, also. This drone brood, which occurs in irregular patches with domelike caps, is often allowed to die. It may be found in various stages of decay, usually in moist, pasty, brown patches having a sweet-rotten odor.

Other problems

Queenless colony. A colony without brood during the active season normally is queenless. Queenless colonies usually become very disturbed when the hive is opened. Scattered cells of pollen with a glossy appearance, found in the area normally occupied by brood, is the typical indication of a queenless colony.

Dysentery. This is a functional disorder that may result from eating indigestible food during a prolonged period of confinement. The most noticeable sign of dysentery is fecal matter in the hive or around the entrance, as bees normally void their body wastes while in flight outside the hive. Honeydew, unripened honey, overheated honeys, or fermenting sugar syrup are unsuitable as feed and will cause dysentery if bees are unable to make frequent flights to void body wastes (as is common in winter).

Starvation. Starvation is a major cause of colony loss in winter and spring, but can occur at any time of the year. Conclusive proof of starvation is the presence of clusters of dead bees, stuck head first in comb cells where they have died in search of food. Often, colonies that have produced a large amount of early brood deplete their stores and die of starvation during confining weather in spring. Colonies also may starve later in the season if subjected to a prolonged dearth between nectar flows.

The most obvious sign of approaching starvation is loss of hive weight or absence of sealed honey during a nectar dearth. Starving bees are restless and crawl slowly about the comb as though cold, even though the weather may be warm. Egg laying is retarded or ceases entirely, and brood is neglected and allowed to die. Cappings sometimes are removed and the brood eaten. Starving bees occasionally cluster in a hunger swarm, usually on or near the hive.

Pests of Bees

Wax moth

The greater wax moth, *Galleria mellonella* L., occurs in all areas of California and is active year-round in coastal and southern California. Night-flying female moths (Plate III) lay eggs inside and outside beehives, showing a preference for stronger colonies. Wax moth larvae tunnel through the combs chewing up the wax and obtaining nutrients from pollen, cocoons, and other debris. Without control, wax moths can reduce combs to a mass of webbing and fecal pellets.

Fortunately, honey bees are capable of locating and removing wax moth larvae before they do much damage, as long as the colony is strong and is not given too much space to patrol. Any supers of combs brought in from the field for storage are very likely to be infested. If a good-sized deep freeze is available, exposure to freezing temperatures for a day or two will kill all life stages of the moth. Otherwise, beekeepers rely on fumigation with various chemicals to kill wax moths. (Chemicals and biological control agents currently registered for wax moth control and their use are explained in the box, *Materials Registered for Wax Moth Control*.)

Other, smaller moths, commonly encountered as pantry pests, will invade stored honeycombs, also. However, these moth larvae usually cause much less damage to the comb and are eliminated by most treatments effective for greater wax moths (but not Certan).

Ants

Normally, a stong, healthy colony of bees can repel an attack by most types of ants. However, in California, Argentine ants are capable of destroying nuclei and full-strength colonies. Often, the ants begin foraging on nectar and honey. Lines of marching ants can be seen leading to the hole they use to enter and leave the hive. Before long the bees become demoralized by the ants and fail to forage properly. This is detrimental to the colony and to the grower, if the bees are being rented for pollination. Left unchecked, the ants often move their headquarters directly below the hive, continue to remove honey and pollen, and eventually begin eating the brood. The bee colony will perish.

Beekeepers use two approaches to ant control: physical barriers and insecticides. Where practical, hives can be placed on low stands that have a can of water or oil at the base of each leg. Water evaporates quickly without a thin layer of oil on it. When cans go dry, become filled with floating dust or debris, or are circumvented by a grass-blade bridge to the hive, the ants will be back in no time.

Various forms of ant baits are on the market in small individual containers and are more or less successful, depending upon what else is available to the foraging ants. A number of insecticides are labeled for controlling ants on the ground, but remember that bees are susceptible to the chemicals, also. Be sure to use insecticides cautiously around bees. (See also *An Observation Beehive*.)

Bears

Insects make up a substantial portion of a bear's normal diet, so it is not surprising that bears will tear open beehives, wild or human-made, to seek out the brood for food. Bears also like honey.

Beekeepers can protect their hives adequately with well-constructed

MATERIALS REGISTERED FOR WAX MOTH CONTROL

Certan. This is a wax moth-adapted strain of *Bacillus thuringiensis* that can be sprayed on combs. Coverage must be thorough and older larvae are much less susceptible to the bacterium than younger larvae. Bacterial spores should be effective for 12 months if temperatures remain reasonable. Bees clean the spores off the combs and contamination of honey is no problem.

Aluminum phosphide. A fumigant formulated as a slow-release, dry, solid pellet or tablet. Easy to use in a reasonably moist environment, it provides excellent penetration of supers and combs, killing all life stages of the moth, if fumigant concentration is held at adequate levels long enough. The highly volatile gas escapes quickly from tiny exit holes in fumigation chambers and has given poor results in a number of warehouses that were not well sealed.

Paradichlorobenzene (PDB). Effective on larval and adult stages, but eggs survive. Covered stacks of five or six deep supers or ten shallow supers (empties, only) are treated with 3 ounces of crystals. At least two treatments at 4- to 21-day intervals are needed to eliminate newly hatched larvae. Combs must be "aired" adequately before use to prevent problems with bees.

electric fences in areas where native bears are known to roam. However, once a bear has become accustomed to destroying hives, it takes much more than a fence to eliminate the habit. It is much easier to inquire about bears at the local county offices before moving into an area than to try to deal with the problem later.

Skunks

It does not take a skunk long to learn that bees crawl out of a hive entrance if the hive is scratched at night. The bees become tangled in the skunk's hair, and the skunk eats them. Toenail scratches on a hive or, sometimes, a hole being excavated under a side of the hive indicate skunk activity. Repeated visits to the same hive can detrimentally affect the population and colony morale. The latter effect will become very evident to the beekeeper when the hive is opened! Skunks can be trapped, but use of poisons for skunk control in California is illegal.

Mice

As the weather becomes cooler in fall, field mice seek protected places to overwinter. Stacks of stored bee combs, or even the lower, empty boxes of combs in hives from which the cluster has moved up, are attractive to mice. A mouse nest usually involves four or more combs hollowed out to support dried grass, leaves, and bits of cloth used to build the nest. By spring, the family has increased in numbers and the odor has become unacceptable.

Entrance guards of 3/8-inch mesh hardware cloth will protect hives in the field, and tight-fitting pallets and covers should protect stored combs.

Livestock

Occasionally, cows and other livestock may push a hive to see what it is or they will use it to scratch an itch. Normally, the hives are disturbed only once. It is best to avoid direct contact with horses since they react violently to beestings, prompting more stings and leading in some cases to severe self-injury.

Vandalism

Vandalism often is avoided by placing the apiary in clear view of passersby and by posting a notice to the effect that witnesses to vandalism will receive a reward for supplying evidence leading to the vandal's prosecution. Various California beekeeping organizations provide such posters to their members.

An Observation Beehive

Few hobbies are as exciting and as educational as keeping a colony of bees in an observation hive. The behavior of bees as a social unit and their elaborate means of communication can be used to illustrate basic biological concepts in teaching at all levels from kindergarten to college.

Once the colony is established behind glass walls, anyone can visually enter the world of the honey bee to observe the activities of an intriguing society. All of the honey bee's life in the hive is unveiled, from egg laying by the queen to the emergence of newborn worker bees from cells in the comb. Some bees will be seen processing pollen into bee bread, others will be converting nectar into honey, and many other worker bee activities, such as cleaning the nest, building comb, and exchanging food, can be seen night and day. But perhaps the most fascinating sight is that of the worker bees returning from foraging trips heavily laden with brightly colored pollen pellets as they enter the hive and perform dances that tell other workers where the pollen was gathered. (As an added bonus, bees in a four-frame observation hive may produce up to 20 pounds of honey annually.)

To fully appreciate an observation hive, you should have some background knowledge about the biology and behavior of honey bees. Many excellent books on bees are available, and a few hours of reading some will pay great dividends.

The hive itself should be located so as to permit observation from both sides. Access must be provided from the hive to the outdoors so that the bees can forage for food and water. A transparent runway through the wall or a window will provide for this. Ideally, there should be no sidewalks or parking areas within approximately 30 feet of the exit. Runways can be of considerable length and can be built to turn corners or curves, although bees seem to orient better if they can see light at the runway's exit.

Construction and mounting (fig. 14)

Observation hives can be purchased, but the hive described in this text is economical and simple to construct and will accommodate four standard full-super frames. Bees need this amount of space for clustering, rearing brood, and storing food reserves.

Ideally, the hive base should be mounted rigidly to a sturdy table or platform. Because all manipulations of the colony must be made outdoors, the mounting and runway attachments should be made so that the hive can be disconnected easily. Before the bees are installed, temporarily mount the hive in its permanent position and then construct the runway to the outdoors. Runways can be made with parallel wood strips on a wooden floor and covered with glass, Plexiglas, or plastic. Fibrous material, such as cardboard, paper, or cloth, should never be used; bees chew through these materials in a few days.

Sometimes there is a problem in making an opening through the window to accommodate the exit runway. One solution is to replace the window glass with a sheet of Plexiglas or plywood in which an opening can easily be cut. For an attractive installation, paint all wood parts (except the frames) of the hive and runway before the glass or plastic is mounted. White observation hives are most attractive, but any color is satisfactory. Paint should be dry before bees are placed in the hive.

Establishing the colony

Bees may be installed in the observation hive anytime between

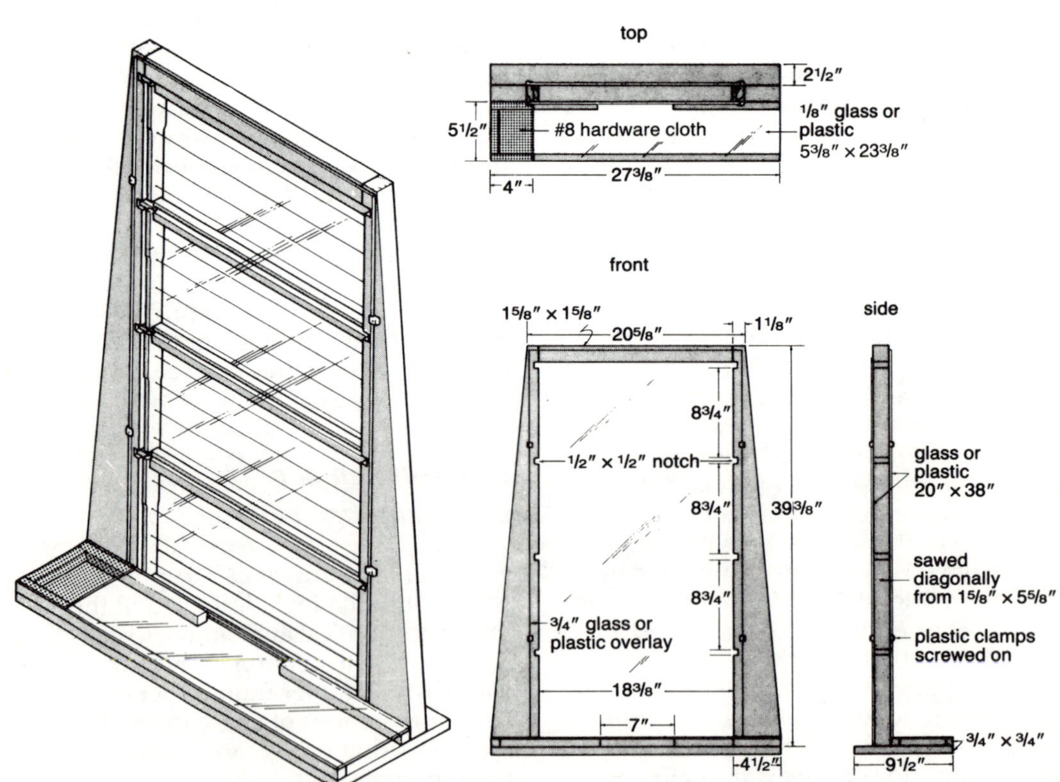

Fig. 14 An observation hive, with construction details. Commercial observation hives are available, also.

early spring and midsummer. Worker bees may be purchased along with the queen; approximately 3 pounds of bees are sufficient.

The quickest method to establish an observation hive is to put frames of brood and a queen from a conventional hive into it. Once the queen is inside the observation hive, the temporarily disorganized bees (including those outside the hive) will soon find the queen and cluster around her.

Instead of purchasing packaged bees, a swarm may be captured and installed. During the swarming season each spring, various public agencies (police, fire department, county agricultural agencies) receive numerous requests to remove swarms, and they frequently are willing to place applicants' names on a "swarm waiting list."

Installing a swarm. Lay the observation hive (containing frames) on its side with the runway side up, propping the top of the hive on a box approximately 1 foot high. Loosen the plastic mounting clamps on the upper glass wall and slide the glass approximately 1 foot toward the hive top. Then shake the cluster of bees into the opening and gently slide the glass wall into position, being careful to avoid crushing bees. Inevitably, a few bees will not get into the hive, and these should be checked to see if the queen is among them. If the queen is among them, she should be captured and placed in the hive.

Installing packaged bees. Prepare the hive as in the instructions immediately above. Now, lightly sprinkle water on the wires of the package — this will calm the bees. Rap the package so that worker bees will fall to the bottom, and then remove the queen cage from the package. One end of the queen cage has a hole with a cork disk over it; remove this disk, exposing the candy beneath it. Place the cage inside the hive near the lower frame, making sure that the cage's screen can be reached by worker bees (they will have to feed the queen through the screen for a few days).

Now shake the bees into the hive and slide the glass wall shut. The bees will be attracted to the queen and will eat the candy that blocks her exit from the queen cage, thus freeing her. If the cage is not supplied with candy, the queen should be released immediately. The empty cage can be removed when convenient.

Transferring bees from conventional hive to observation hive. Remove two frames of capped brood, one frame of honey, and one frame of empty comb from a conventional hive (all frames should be covered with bees). Place them in that order, bottom to top, in the observation hive. Shake additional bees from the conventional colony into the observation hive. Make certain that the queen has been transferred.

Maintaining the hive

After the newly established hive is mounted, a feeder containing sugar syrup should be provided for the colony. Feeders can be made by punching or drilling 20 to 50 small holes in the lid of a pint or quart glass jar; the jar should then be filled with sugar syrup and inverted over the feeding chamber. Sugar syrup should be made available continuously until all the combs are filled with honey or brood. Thereafter, the colony should be fed only when its stored honey is gone.

Under normal conditions established colonies are self-supporting and require little maintenance. However, colonies in observation hives require special maintenance because there are fewer foragers than in the regular hive. When weather conditions permit foraging flights, and nectar and pollen are available, the observation colony collects nectar rapidly and accumulates an abundance of honey, which reduces the need for maintenance.

Preparing the colony for winter. Unless the climate permits bee flight at least once a month, it is not advisable to try to maintain an observation colony in winter. Without periodic flights, high mortality usually occurs, and the colony may die in midwinter or early spring. Therefore, it is usually best to terminate the colony in autumn after brood rearing has ceased (the queen can be removed earlier if desired). This is done by shaking the bees off the observation hive frames near the entrance of a normal outdoor colony. The bees will soon be accepted into the colony. The frames of combs from the beeless hive may then be wrapped and stored at 0°F; this prevents granulation of honey and infestation by pests during storage. The following spring a colony may be reestablished in the hive, using the stored frames of comb.

Problems and solutions

Although honey bees are largely self-sufficient, minor difficulties may arise occasionally. These are discussed below.

Sunlight. Observation hives should never be exposed to direct sunlight.

Ventilation. Normally, the observation hive will have adequate ventilation through its runway to the outside and additional ventilation ports will not be necessary. However, if the inside of the hive walls becomes fogged for a prolonged period, additional ventilation ports (3/4-inch holes covered by 8-mesh wire screen) may be provided on the top or ends of the hives. Healthy colonies typically are full of bees, and it is a mistake to suppose that bees need additional ventilation simply because they appear to be crowded.

Swarming. In spring colonies increase rapidly in population, and swarming is therefore to be expected. Hobbyists may wish to study this phenomenon, but if they wish to prevent it, the easiest control is to kill the old queen (by pinching her head) when the colony population reaches its peak in spring. A new queen will be reared automatically by the bees, and the short interruption of brood rearing normally stops swarming tendencies for the remainder of the season.

Invasion by pests. In some areas ants are serious pests of bees; colonies invaded by ants are liable to become disorganized enough to stop normal activities. Poisonous baits for ant control may be used near the colony, but access by bees (or other animals) to baits must be prevented by covering bait containers with 8-mesh wire screens, which should be at least 1/2 inch from the bait itself so that bees cannot reach through and eat the bait. Do not use insecticides near the hive.

Population decline. Except for normal seasonal fluctuations, a decline in bee population usually is caused by insufficient brood rearing. Usually, the hive population is stable; hundreds of new bees emerge each day and compensate for normal losses (bees live 6 to 8 weeks in summer and up to 6 months in winter). If brood-rearing decline is caused by an old and inferior queen, replacing the queen is usually the best solution (see requeening in *Maintaining Genetic Stock*).

Lack of food. The threat of starvation is greatest when rapid consumption of hive food supplies occurs during the intensive spring brood rearing. If the hive contains enough capped cells of honey, bees will not starve. If capped honey is not present, sugar syrup must be fed to the colony.

Accidental bee escapes. Because they are confused, bees accidentally released indoors usually do not sting. However, stinging may occur near the colony within a few seconds after bees escape, particularly if thousands are liberated suddenly. If this happens, permit the colony to settle down for a few minutes. After the bees have become settled, the hive and any adhering bees may be gently taken outdoors. (Any bees remaining in the building may be caught easily with a vacuum cleaner.) Whenever the colony is carried outdoors, always remember to plug up the runway at the point where it is disconnected from the hive.

Orientation of bees. Observation hive bees can become disorganized (disoriented) when they are installed, or after any change in the arrangement of the colony runway. Disoriented bees in a hive seem to be wandering about and do not perform any of the chores they usually do. Several days may be required for forager bees to adjust to a new location or runway arrangement. Young bees just learning to fly may be seen in intensive flight around the hive entrance in early afternoons; this is their method of orienting themselves to the colony in preparation for later foraging.

Use of smoker and protective clothing. To control bees, a few gentle puffs of smoke should be blown into the hive entrance just before the top of the hive is removed. When smoke is applied skillfully and in small amounts, the risk of being stung is minimized; however, one should always move slowly and carefully around bees — fast motion, strong vibrations, or any jarring of the hive excites them.

Glossary

Abate
To eliminate a (disease) problem by removing (often by burning) or treating bees and beekeeping equipment so that there is no possibility of contaminating other bees.

Acid board (*also *Fume board)
A rimmed hive cover containing a pad of absorbent material into which benzadehyde or butyric anhydride (bee repellents) is poured. Used to remove bees from honey supers.

Apiary
A collection of one or more populated beehives at a certain location.

Bee bread
Bitter, yellowish pollen stored in honeycomb cells and used by bees for food.

Bee escape
A mechanical device that allows bees to pass through it in only one direction. Often a leaf spring or cone design used to eliminate bees from particular supers in a hive or from buildings.

Bee glue
See *Propolis.*

Beehive
Normally refers to a human-made container in which the colony lives. Movable frame hives are required by law in California (see *Hive*).

Beekeeper
An individual who oversees the maintenance of one or more colonies of bees.

Beesting
The apparatus at the tip of an adult female bee that can inject venom into the victim being stung. The worker sting remains in the victim and continues to inject venom; it should be scraped off sting site.

Beeswax
Wax secreted by glands located on the underside of four abdominal segments of the honey bee. It is used by bees to construct comb.

Boardman feeder
A small, wooden feeder placed at the hive entrance and holding an inverted pint or quart glass jar of sugar syrup. Not recommended.

Brood
Any immature stage of development: egg, larva, or pupa. Also, collectively, all immature bees in the hive.

Brood comb
Any drawn comb in which eggs, larvae, or pupae are found.

Brood nest
The area inside the hive body devoted to brood rearing.

Brood rearing
The process involving egg laying, feeding larvae, and keeping pupae warm, which produces more adult bees.

Cappings
A thin layer of wax covering ripened honey or developing pupae. Cappings are collected when honey is being uncapped. Capped brood refers to pupae.

Cappings melter
A hot water, steam, or electrically heated container used to separate honey and wax by melting; wax floats on the honey.

Cappings spinner
A centrifuge with wire-screened baskets used to separate honey from wax.

Cell
One of the hexagonal compartments of a honeycomb in which brood is reared or food is stored.

Cismontane
Area west of Sierra Nevada Mountains in northern and central California, and area west of Mojave and Colorado deserts in southern California. (See also *Transmontane.*)

Clipping and marking
Terminology referring to the clipping of a portion of a queen's wings and the affixing of a dot of colored material on the top of her thorax.

Cluster
Loosely, any group of bees that forms a relatively compact aggregation. A winter cluster is composed of all the bees in the colony huddled as closely together as necessary to maintain the required temperature. As the ambient temperature increases, the cluster expands until it loses its identity but it will reappear if the temperature drops.

Colony
A community of bees living in close association and contributing to their mutual support by their labor. It is composed of a queen and worker bees, and during spring and summer drone bees are present. The terms colony and hive are often used interchangeably.

Comb
A mass of hexagonal cells made of beeswax and containing brood and food.

Cover (also referred to as a top or lid)
The flat, wooden piece placed on top of the hive to confine and protect the bees.

Crosspollination
Movement of pollen between blossoms of one variety of plant species and a second, compatible variety to produce hybrid seed. (See also *Pollination*.)

Dearth
Severe to total lack of availability, usually in reference to nectar and/or pollen.

Demaree method
A swarm prevention technique based on removal and isolation of a colony's brood at the top of a multiple-story hive.

Drift
Movement of bees from their original hive into a neighboring hive—frequent with drones and surprisingly common with workers.

Drone
A male bee that develops from an unfertilized egg.

Dysentery
Intestinal disorder causing frequent defecation (diarrhea) in affected individuals. Tan, brown, or black fecal smears on combs or outside of hive indicate such a problem.

Escape board (also, sometimes, inner cover)
A device with dimensions identical to the top of a super that contains one or more bee escapes. Used to empty one or more supers of bees.

Extractor
A mechanical device used to remove honey from uncapped honeycombs by centrifugal force.

Festoon
A unique cluster of bees that link themselves together by their tarsi (feet) in a loose network between combs in a hive. Normally, these are aggregates of wax-producing bees.

Flow
Refers to the availability of nectar and/or pollen. When food substances are available in abundance, it is a "good flow."

Foraging
Those activities of bees connected with finding and bringing back water, nectar, pollen, or propolis.

Foundation
A thin sheet of beeswax imprinted with the hexagonal cell bases of a honeycomb; used as a base for the comb when placed in frames.

Frame
A rectangle, usually of wood, that is hung inside the hive to support the foundation and comb. Sometimes frame and comb are used interchangeably; that is, a "comb of brood" is a "frame of brood."

Fume board
See ***Acid board.***

Hive
A container housing a colony of bees. Usually consists of one or more hive bodies below and one or more supers above. (See ***Beehive*** and ***Colony***.)

Hive body
The part of the hive containing combs in which the queen lays eggs. The hive body rests on the bottom board.

Hive stand
A device that elevates the bottom board up off the ground.

Hot room
An insulated portion of a warehouse with radiant or forced air heating that can produce temperatures up to 100°F.

Larva
The wormlike immature stage of a honey bee that increases in size dramatically as it feeds on royal jelly, pollen, and diluted honey.

Nectar
A dilute sugar solution secreted by glands in different parts of plants, chiefly in flowers.

Nuclei
A small functioning colony of bees (queen, bees, brood) on two to five combs.

Nurse bee
A worker bee of the correct age (6 to 12 days postemergence) to produce royal jelly and to feed larval bees, adult queens, and drones.

Oven
A small, highly insulated portion of a warehouse, often in the hot room, where temperatures can be elevated to 150°F to melt wax.

Package
A wire-screened wooden box of bulk bees, a queen, and a can of feed used to transport bees to an empty hive.

Pollen
Male sex cells produced in anthers of flowers. Powderlike and composed of many grains, they are gathered and used by honey bees for food as a source of protein. A good mix of many different pollens is essential for adequate nutrition.

Pollination
Transfer of viable pollen to a receptive stigma of a flower. In commercial beekeeping, the term refers to the service provided by honey bees in crop production. (See also *Crosspollination*.)

Pollen substitutes
Feed substances fed to bees to provide protein, fats, vitamins, and minerals when pollens are not available.

Pollen supplement
Pollen substitute mixed with pollen to increase attractiveness and nutritive value to bees.

Pollen trap
A device attached to a hive to remove pollen loads from incoming foraging bees. Pollen "pellets" usually are collected in a drawer that is inaccessible to the bees.

Prepupa
An immature stage between the last larval stage and the true pupal stage in the life cycle of a honey bee.

Propolis
Plant resins collected by bees and used as a cement to stick hive parts together and to seal openings. Also called bee glue.

Pupa
The preadult form of bees occurring after the larval stage and maintained without evident change in size and structure until the adult bee emerges from the cell.

Queen
Lone, fully developed female in colony. She lays all the eggs and stores sperm for up to 3 years.

Queen cage candy
A special fondant made from Nulomoline, drivert, and glycerine (see *Feeding Bees*); used to feed queen and attendant bees in queen cages.

Queen excluder
A wire or plastic grid, with slots just large enough for passage of worker bees, used to prohibit the movement of queens between supers.

Queenless
A hive of bees with no queen.

Queenright
A colony of bees with a functioning queen.

Rendered comb
Comb that has been melted down to beeswax. With American foulbrood, the wooden frames are soaked in a lye bath.

Requeen
To remove the present queen from the colony and replace her with another queen.

Ropiness
Having the characteristic of sticky elasticity and stringing out when stirred and stretched.

Royal jelly
A glandular secretion from the heads of worker bees used to feed young larvae and adult worker, drone, and queen bees.

Scale
A dehydrated, dead larva shrunken to an elongated thin, flat chip at the bottom of a cell.

Slumgum
A mixture of propolis, pollen, cocoons, and other debris that persists after beeswax and honey have been recovered from rendered combs.

Solar melter
A device designed to use the heat of the sun to melt beeswax, and, in some cases, to separate honey from beeswax.

Spermatheca
A small, round organ in the abdomen of a queen bee capable of storing viable sperm for 3 years.

Spring dwindling
A condition in which the colony population decreases in size during spring at which time exponential population growth is anticipated.

Super
A wooden box with frames containing foundation or drawn comb in which honey is to be produced. Named for its position above the brood nest. The same type of box is referred to as a hive body when it is situated below the honey supers and is intended to be used for brood rearing and pollen storage.

Supersedure
A natural process by which a colony of bees replaces its present queen with a new one.

Swarm
A cluster of worker bees, with or without drones and a queen, that has left the hive.

Trachea
A system of air-filled branching tubes that conduct oxygen from outside the body to inner tissues of the bees.

Transmontane
Area east of Sierra Nevada Mountains; includes Mojave and Colorado deserts.

Wintering
The process of preparing the hive and colony for survival over winter. Also, a colony in the process of attempting to survive over winter.

Worker
An infertile, female honey bee, anatomically adapted to perform the work for a colony of bees including: manipulating stored food, feeding brood, guarding hives, foraging for food, etc.

References

Many books have been written on beekeeping. Generally, the larger and more expensive the book, the more comprehensive the information. This list includes only a few representative books by category, but many others are available through bookstores and beekeeping supply dealers. Many good pamphlets are available, also, from the county offices of UC Cooperative Extension.

Title	Author	Publisher	Region of coverage	Approx. price
How To				
Begin to Keep Bees	Carrier	Carrier	Western U.S.	$15
First Lessons in Beekeeping	Dadant	Dadant	Eastern U.S.	$ 2
How to Keep Bees and Sell Honey	Kelley	Kelley	Eastern U.S.	$ 2
Mastering the Art of Beekeeping	Aebi	Rodale	Western U.S.	$ 8
Starting Right with Bees	Gleanings	Root	Eastern U.S.	$ 2
The Art and Adventure of Beekeeping	Aebi	Rodale	Western U.S.	$ 5
Comprehensive Texts				
Bees and Beekeeping	Morse	Comstock	Eastern U.S.	$30
Bees, Beekeeping, Honey and Pollination	Gomjerac	AVI	Eastern U.S.	$20
The Hive and the Honey Bee*	Grout	Dadant	Eastern U.S.	$15
Reference Books				
ABC & XYZ of Bee Culture	Root	Root	Worldwide	$14
A Scanning Electron Microscope Atlas of the Honey Bee	Erickson	Root	Worldwide	$50
Honey—A Comprehensive Survey	Crane	IBRA	Worldwide	$28
Honey Bee Pests, Predators and Diseases	Morse	Comstock	Worldwide	$43
The Illustrated Encyclopedia of Beekeeping	Morse & Hooper	Root	Worldwide	$35
Special Topics				
Contemporary Queen Rearing	Laidlaw	Dadant	U.S. practices	$11
Honey in the Comb	Killion	Dadant	Eastern U.S.	$10
Instrumental Insemination	Laidlaw	Dadant	U.S. practices	$14
Making Mead	Morse	Scribner	Worldwide	$10

*Best comprehensive text available.

Beekeeping periodicals

Beekeeping periodicals provide current information on many aspects of the industry. They also contain a wealth of advertising. The following list includes the major, English language periodicals with their areas of emphasis. Check with the Extension apiculturist to determine whether the state is still publishing a beekeeping newsletter.

American Bee Journal, Hamilton, IL 62341. Emphasis on concerns of the commercial industry, research, and some how-to-do-it information.

Gleanings in Bee Culture, P.O. Box 706, Medina, OH 44258-0706. Emphasis on how to do it, with information on research and concerns of the commercial industry.

The Speedy Bee, P.O. Box 998, Jesup, GA 31545. Newspaper format with emphasis on federal and state governmental actions concerning beekeeping. Research results and specific management techniques sometimes included.

International Bee Research Association, 18 North Road, Cardiff CF1 3DY, United Kingdom (England). The world's only organization devoted to collecting and disseminating beekeeping information globally. Publishes three English language journals:

Apicultural Abstracts—English language synopsis of every available article containing information on bees around the world.

Bee World—Excellent review articles and news briefs.

Journal of Apicultural Research—Current research.

Cooperative Extension publications

The following priced publications about beekeeping may be obtained by writing ANR Publications, University of California, 6701 San Pablo Avenue, Oakland, CA 94608-1239. Ask for the Catalog that lists the prices of each publication listed.

American Foulbrood Disease (Afb) of Honey Bees (2757)
Identification, causes, control, and prevention.

Bee Problems in Outside Dining Areas (2852)
How to eliminate them.

Bee-ginner Beekeepers (2764)
Responsibilities and equipment involved in beekeeping, instructional resources available, and sources of beekeeping supplies.

Economic Trends in the U.S. Honey Industry (21219)
Published in 1980.

Honey Bees in Alfalfa Pollination (2382)

Honey Bees in Almond Pollination (2465)
Factors affecting pollination, ways to maximize bee pollination, sample contract for growers and beekeepers.

Honey Bee Pollination of Cantaloupe, Cucumber, and Watermelon (2253)
How to manage honey bees for effective pollination.

How to Construct and Maintain an Observation Beehive (2853)
Plans for a glass-walled indoor observation hive, for teaching, recreational, or scientific use.

Making and Using a Solar Wax Melter (2788)

Reducing Pesticide Hazards to Honey Bees with Integrated Management Strategies (2883)
Applicable to forests, rangelands, recreational and residential settings, and agricultural crops.